LIE GROUPS: HISTORY, FRONTIERS AND APPLICATIONS

VOLUME V

SYMPLECTIC GEOMETRY AND FOURIER ANALYSIS

NOLAN R. WALLACH

Department of Mathematics
Rutgers University
New Brunswick, New Jersey 08903

WITH AN APPENDIX ON QUANTUM MECHANICS

BY ROBERT HERMANN

MATH SCI PRESS

53 JORDAN ROAD

BROOKLINE, MASSACHUSETTS 02146

LIE GROUPS: HISTORY, FRONTIERS AND APPLICATIONS

VOLUME V

SYMPLECTIC GEOMETRY AND FOURIER ANALYSIS

Library of Congress Catalog Card Number: 76-43503
ISBN: 0-915692-15-5

MATH SCI PRESS
53 JORDAN ROAD
BROOKLINE, MASSACHUSETTS 02146

Printed in the United States of America

PREFACE

Most mathematicians (myself included) are awed and somewhat mystified by what physicists call quantum mechanics. Most recent books by physicists on quantum mechanics begin with a mathematical framework (generally Hilbert space theory). They develop standard results from functional analysis using very suggestive terminology. They then set up some framework that leads to the Schrödinger equation. These equations are usually studied by separation of variables. At some point in this development the mathematics undergoes a "transformation into physics". The mathematical rigor dilutes and the concepts become progressively more foreign to mathematicians. There is no classical analogue of the mathematical softness of quantum mechanics. Classical mechanics begins with some variant of Newton's laws. The necessary mathematics consists of differential equations. In general, for concrete problems, the differential equations are hard to solve. A variety of variational and geometric alternative interpretations are made. The mathematics in these schemes becomes progressively more sophisticated. When classical mechanics leaves the realm of mathematical rigor, it does so for purely pragmatic reasons: The mathematical theorems are not sufficient to handle the problem. This route to physics is (in my mind) in no way mysterious to a mathematician.

Recently, many mathematicians have attempted to make the route to the Schrödinger equation more satisfactory to mathematicians. The mathematical motivation for this work comes from the great success of Kirillov's technique for studying harmonic analysis on nilpotent Lie groups, and Auslander-Kostant's sweeping extension of the theory to solvable groups. These mathematical theories begin with the Stone-Von Neumann theorem, which in a sense makes rigorous the simplest example of a route to the Schrödinger equation: the Heisenberg quantization rules for a free particle.

The work of Auslander-Kostant is an outgrowth of Kirillov's theory and Kostant's theory of quantization (this theory has been studied most significantly by Kostant, Blattner [3], Sternberg and Guilliman). The Kostant quantization is mathematically rigorous. However, it replaces the "physical mystery" by a very disheartening mathematical fact: It is not always possible to carry out the quantization rules. Also, even when it is formally clear that the quantization rules can be carried out, the actual process leads to (for example) divergent (singular) integrals which must be interpreted in some renormalized sense (i.e., through analytic continuation). An interesting example of this type comes from replacing the Heisenberg group by $SL(2, \mathbb{R})$. Then the Kostant quantization involves the Kunze-Stein intertwining operators.

However, Kostant's theory has an important redeeming feature. There are two steps in the procedure of quantization. The first step is pre-quantization. Pre-quantization is always possible. It also leads to interesting mathematics. Quantization becomes a method of "cutting down" the number of variables involved in pre-quantization. For example, in the Heisenberg quantization rules there are two steps. Let us look at this example in a bit of detail. In classical mechanics one labels a particle at a time t by its position $q(t)$ and its momentum $mq'(t) = p(t)$. Thus one looks at a particle as a curve in $\mathbb{R}^3 \times \mathbb{R}^3$. Newton's laws (if the force law is conservative) in Hamiltonian form become

$$\frac{dq_i}{dt} = \frac{\partial H}{\partial p_i}(q(t), p(t))$$

$$\frac{dp_i}{dt} = -\frac{\partial H}{\partial q_i}(q(t), p(t))$$

where H is the Hamiltonian of the system. If f is a function on $\mathbb{R}^3 \times \mathbb{R}^3$ (say of class C^∞) then

$$\frac{d}{dt} f(q(t), p(t)) = \sum \frac{\partial H}{\partial p_i}\frac{\partial f}{\partial q_i} - \sum \frac{\partial H}{\partial q_i}\frac{\partial f}{\partial p_i}$$

$$= \{H,f\}(q(t), p(t)) .$$

$\{H,f\}$ is given by the above formula and is called the Poisson bracket of H with f. Now, if we set $U_t(q,p) = (p(t), q(t))$, where $q(0) = q$, $p(0) = p$, then $U_{t+s}(q,p) = U_t(U_s(q,p))$ by uniqueness. This gives

$$\frac{d}{dt} f(U_t(q,p)) = \{H,f\}(U_t(q,p)) \quad .$$

Thus we can think of the trajectories of the particles in the Hamiltonian field as the trajectories of the one parameter group generated by the vector field $f \to \{H,f\}$. We have therefore reinterpreted the Hamiltonian H as an operator on the C^∞ functions on $\mathbb{R}^3 \times \mathbb{R}^3$. The trajectories of the particles have become a one-parameter group of transformations of $C^\infty(\mathbb{R}^3 \times \mathbb{R}^3)$. That is, $(U_t f)(x) = f(U_t x)$. The equations of motion are

$$\frac{d}{dt} (U_t f) = U_t\{H,f\} \ .$$

The first step in pre-quantization of this problem is to add a complex variable. That is, look at f as "its function value". In other words, $x \to (x, f(x))$ becomes a function from $\mathbb{R}^3 \times \mathbb{R}^3 \mapsto \mathbb{R}^3 \times \mathbb{R}^3 \times \mathbb{C}$. The next step is to interpret $\omega = \Sigma dp_i \wedge dq_i$ as a curvature. There are many ways of accomplishing this (the fact that they will give the same result is the independence of quantization in this special case); one is to write $\theta = \Sigma p_i dq_i$, then $d\theta = \omega$. If $\phi \in C^\infty(\mathbb{R}^3 \times \mathbb{R}^3)$ then we define

$$(\delta(\phi)\cdot f(x) \quad = \quad (\{\phi,f\}(x) - 2\pi i\theta(x_\phi)f(x) + 2\pi i\phi(x)f(x)) \quad .$$

Here X_ϕ is the vector field given by $X_\phi(f) = -\{\phi,f\}$. δ is called pre-quantization. One finds by a simple computation that

$$(\delta(\phi_1),\delta(\phi_2)) \quad = \quad \delta(\phi_1)\delta(\phi_2) - \delta(\phi_2)\delta(\phi_1)$$

$$= \quad \delta(\{\phi_1,\phi_2\}) \quad .$$

(See Section 2.6.)

If $\phi(q,p) = \Sigma a_i q_i + \Sigma b_i p_i + t$, (that is, a polynomial of degree at most one). Then

$$\delta(\phi) \quad = \quad \sum a_i \frac{\partial}{\partial p_i} - \sum b_i \frac{\partial}{\partial q_i} - 2\pi i \left(\sum b_i p_i + t\right) \quad .$$

For these functions we can "cut down" the number of variables in the domain of $\delta(\phi)$ by the obvious trick. Let

$$H_\infty \quad = \quad \left\{ f \in C^\infty(\mathbb{R}^3 \times \mathbb{R}^3) \,\middle|\, \frac{\partial f}{\partial q_i} = 0,\ i = 1,2,3 \right\} \quad .$$

If $f \in H_\infty$ and ϕ is as above, then

$$\delta(\phi)f \quad = \sum a_i \frac{\partial f}{\partial p_i} - 2\pi i \left(\sum b_i p_i f + t\right) \quad .$$

This is the Heisenberg quantization of linear functions. The "ad hoc" rules $q_i \mapsto \partial/\partial p_i$ and $p_i \mapsto$ multiplication by $-2\pi i\, p_i$ have been incorporated in a formalism. The Stone-Von Neumann theorem now says that subject to certain regularity conditions (essential self adjointness) this is the only way we could have cut our variables down from six to three.

In the above example two properties become apparant. The first is that we can "pre-quantize" any function. To quantize one must use very special functions. We will see in the discussion in Chapters 5 and 6 of the metaplectic representation that the Heisenberg quantization can be extended to polynomials of degree at most two. This gives a the quantization of a free particle in the absence of forces. Once forces are introduced the theory becomes murkier; unless they are of special type, the physicist must go back to his old ways.

This monograph is the lightly edited notes for a course that I gave at Rutgers University during the spring semester of 1975 on symplectic geometry and Fourier analysis. The participants in the course (graduate students and some professors) had mixed backgrounds. For this reason I tried to develop the material of the course from the most elementary perspective. That is, the prerequisites of the course were the basic material in a normal first year graduate program, including a decent course in differential manifolds and elementary functional analysis. In a course of this type the lecturer is constantly in the situation of making compromises in the degree of generality of the results proved. For example, in Chapter 1 a more satisfactory (from the modern point of view) development of the subject matter would involve principal fiber bundles, general connections, the Weil homomorphism, etc. as in Kobayashi-Nomizu [15]. However, all we need from the above mentioned theory is given (with most proofs) in Chapter 1. A careful reader will object that for the definition of Čech cohomology we essentially refer to Hirzbruch [10]. However, a glance at Hirzbruch's development of Čech cohomology will show that it is completely within the spirit of these notes.

Another such compromise is in the use of functional analysis. The subject matter of functional analysis is certainly one of the most elegant parts of mathematics. I am well aware that many parts of Chapters 4, 5, and 7 could be simplified by using some of the more sophisticated theorems in functional analysis. For example, the proof of Theorem 4.3.14 would be simplified by referring to the Kaplansky density theorem. However, the density theorem would not be used again. I chose to give a "bare hands" proof of the result rather than apply the theorem. An expert in functional analysis should have no trouble cleaning up the material of Sections 4, 5, and 7.

With these disclaimers in mind, I will now give a chapter by chapter description of the material of these notes. As was observed, the first chapter gives certain geometric preliminaries, beginning with a simple-minded discussion of the de Rham theorem and Čech cohomology. I give a very explicit proof of de Rham's theorem in dimensions 0, 1, 2 since we will be explicitly computing Chern classes in Chapters 2 and 3. Section 2 discusses line bundles. Certainly, from a pedagogic point of view, there is no real difference in the basic material on line bundles and the basic material on vector bundles. However, the definition of Chern classes is significantly easier for line bundles, hence I have chosen to study only line bundles. An interested reader could certainly redo the definition and Lemma 2.2 for vector bundles. Section 3 introduces the notion of connection on a line bundle. Connections will be used for two purposes; to compute Chern classes and to "pre-quantize" (Chapter 2). Our development of Section 3 follows the broad lines of the corresponding development in Kostant [16].

Chapter 2 begins (in Section 1) with the definition of basic properties of symplectic manifolds. I also give, in Section 1, the main examples of symplectic manifolds which will be studied in these notes (see 1.4 of Section 1). In Section 2, I give a proof of the Darboux theorem. Although we will not need the theorem, we will need most of the algebraic preliminaries to the proof we give (following Weinstein [31]). Section 4 begins the material of the course. In Section 4 of Chapter 2 we study the relationship between symplectic and Hamiltonian vector fields (see Lemma 4.4). In Section 5, Kostant's notion of pre-quantization is introduced (a similar discussion can be found in Sauriau [34] or Kostant [16]). In Section 6 a discussion is given of the concepts of the chapter in the special case of the cotangent bundle of $\mathbb{R}^n$. The reader should compare this material with the discussion of the Heisenberg quantization in this preface.

The point of Chapter 3 is to see how the concepts of Chapter 2 apply to the action of a Lie group on a symplectic manifold. We give, following Kostant [16], a recipe for all simply connected, homogeneous, symplectic manifolds. It is quite interesting to see that to describe the standard symplectic manifold $\mathbb{R}^n \times \mathbb{R}^n$ with $\omega = \Sigma dx_i \wedge dy_i$ according to our recipe the natural transitive group is not $\mathbb{R}^n \times \mathbb{R}^n$ but it is the $2n+1$ dimensional Heisenberg group. (See Section 3.3 for the general recipe.) Section 3.4 gives the example of $\mathbb{R}^n \times \mathbb{R}^n$ in full detail. In a sense, Section 3.5 is the most important section of this chapter. It gives the Kostant [16] (see also Sauriau [24]) criterion for when the symplectic structure on a homogeneous symplectic manifold is integral. Under this condition one can apply pre-quantization.

In the Appendix to Chapter 3, I show how the material of Chapter 3 can be modified to apply to tori. In the course of the development following the lines of Chapter 3, the Heisenberg group plays a central role in the study of line bundles over tori. We also study complex structures. The Riemann bilinear relations become quite natural in this context. Results on the classical theta functions (c.f. Weil [30]) then "pop-out" as very natural parts of our theory.

Chapter 4 begins the development of the second topic of the title, "Fourier Analysis", Sections 4.1 and 4.2 give a standard development of the Fourier transform and tempered distributions (except possibly for Lemma 2.5). Section 4.3 gives proof of the Stone-Von Neumann theorem following Simon [23]. Section 4.4 uses the Stone-Von Neumann theorem to give a classification of the irreducible unitary representations of the Heisenberg group (of course that is the original point of the theorem). In the Appendix to Chapter 4 I show how to relate the material of Chapter 3 and its Appendix to the material of Cartier's [5] approach to theta functions. It should be noted that the use of holomorphic induction in Cartier [5] is replaced in the Appendix to Chapter 4 by the interpretation in the Appendix to Chapter 3 of the Riemann bilinear relations. This Appendix also contains the germs of a marvelous general theory of polarizations that is being developed by Roger Howe.

Chapter 5 is devoted to a rather complete discussion of the metaplectic representation. We follow the broad lines of Shale's [22] observation that the metaplectic representation is the symplectic analogue of the spin representation (formally one replaces the Grassmann algebra by the symmetric algebra). A very similar development can be found in Igusa [11]. (Many researchers feel that this representation should be called the oscillator representation because of its direct connection with the simple harmonic oscillator.) In Section 5.4 I pursue this idea and show how the theory of the Hermite functions is a natural aspect of the theory of the metaplectic representation. It should be noted in passing that the one parameter group Exp(tJ) of Section 5.6 has kernel $K(x,y,t)$ as in Wiener [33], Section 7.

Chapter 6 is (in my mind) the least satisfactory chapter in these notes. It is a quite speculative attempt to relate the materials of Chapters 2 and 3 with the material of Chapters 4 and 5. I hope that it does not add to the "mystery" alluded to in the beginnings of this preface.

Chapter 7 can be looked upon as an application of the ideas of the previous chapters to homogeneous symplectic manifolds of nilpotent Lie groups. It is what is now called the Kirillov theory. As I indicated in the Introduction, the Kirillov theory is one of the main motivations for the development of the geometric theory in this book. It is my feeling, however, that the material of this chapter is most naturally studied in the context of the previous theory. The chapter draws heavily on the basic work of Dixmier [6], Kirillov [14], and Pukansky [21]. Two appendices are included which give simple minded developments of trace class operators and induced representations.

In reading this preface a reader will see that the ideas of Kostant are basic to the material of this book. I have also drawn freely from the work of Kirillov and the exposition of P. Bernat, et al. [2]. I make no claims of novelty in any of the basic ideas of these notes. I have benefitted greatly from conversations with Robert Hermann (much of the material in these notes can be found in different form in Hermann's work), Roe Goodman (he helped with the Kirillov theory and showed me his technique of studying the metaplectic representation), Roger Howe (for describing to me on many occasions his marvelous ideas about the metaplectic representation) and Kenneth Gross (who introduced me to the metaplectic representation).

TABLE OF CONTENTS

Page

Chapter 1

DIFFERENTIAL GEOMETRY

In this chapter we compile several results from differential geometry which will be used throughout these notes. We advise the reader to use this chapter for reference purposes only. Most of the material is either part of or easily derived from a graduate course in differential geometry using, for example, F. Warner [27] or R. Hermann [9] as a text.

1. THE DE RHAM THEOREM

Let M be a paracompact, connected, C^∞ manifold. We denote by

$$\mathcal{D}^k(M) \qquad\qquad (\text{resp. } \mathcal{D}^k(M,\mathbb{C}))$$

the differential forms (resp. complex valued) differential forms of degree k on M. We will use the notation $\mathcal{D}^k$ if the context M (or $\mathbb{C}$) is clear.

Let $d\colon \mathcal{D}^k \to \mathcal{D}^{k+1}$ be the operation of exterior differentiation. The <u>de Rahm cohomology</u> "<u>groups</u>" of M are defined by

$$H^k_d(M;K) = \ker d\Big|_{\mathcal{D}^k} / d\mathcal{D}^{k-1} \qquad (K = \mathbb{R} \text{ or } \mathbb{C})$$

$k = 0,1,2,\ldots$, $\mathcal{D}^{-1} = (0)$.

De Rham's theorem relates $H_d^k(M;K)$ with $H^k(M;K)$ where $H^k(M;K)$ is one of the usual cohomology theories on M (we will use the Čech cohomology theory). Since we will need an explicit form of the de Rham theorem in the next few chapters we review in detail the isomorphisms $H_d^k(M;K)$ with $H^k(M;K)$, $k = 0,1,2$. (A complete account of the de Rham theorem can be found in F. Warner [27].)

We recall the definition of the Čech cohomology groups. Let G be an abelian group written additively. If

$$\mathcal{U} = \{U_\alpha\}_{\alpha \varepsilon I}$$

is an open covering of M, then a $\mathcal{U}$-k-cochain on M with values in G is a rule, c, that assigns to each collection

$$U_{\alpha_0}, \ldots, U_{\alpha_k}$$

of elements in $\mathcal{U}$ such that

$$U_{\alpha_0} \cap \cdots \cap U_{\alpha_k} \neq \emptyset$$

an element $c_{\alpha_0,\ldots,\alpha_k} \varepsilon G$. Let $C^k(\mathcal{U};G)$ be the set of all $\mathcal{U}$-k-cochains with values in G. If $c,f \varepsilon C^k(\mathcal{U};G)$ then

$$(c + f)_{\alpha_0,\ldots,\alpha_k} = c_{\alpha_0,\ldots,\alpha_k} + f_{\alpha_0,\ldots,\alpha_k}$$

defines an abelian group structure on $C^k(\mathcal{U};G)$.

Define $\delta\colon C^k(\mathcal{U};G) \to C^{k+1}(\mathcal{U};G)$ by

$$(\delta c)_{\alpha_0,\ldots,\alpha_{k+1}} = \sum_{i=0}^{k+1} (-1)^i \, c_{\alpha_0,\ldots,\hat{\alpha}_{ki},\ldots,\alpha_{k+1}}$$

(here the "roof" over an index means delete the index). It is easy to see that $\delta^2 = 0$. Set

$$H^k(\mathcal{U};G) = \ker \delta\Big|_{C^k(\mathcal{U};G)} / \delta C^{k-1}(\mathcal{U};G)$$

$h = 0,\ldots$ $(C^{-1}(\mathcal{U};G) = (0))$.

Let $\mathcal{U}$ and $\mathcal{V}$ be open coverings of M. Then we use the notation:

$$\mathcal{U} > \mathcal{V} \quad \text{if } \mathcal{U} \text{ refines } \mathcal{V} .$$

If $\mathcal{U} > \mathcal{V}$, choose $\tau\colon \mathcal{U} \to \mathcal{V}$ so that $\tau U \subset U$. If $c \in C^k(\mathcal{V};G)$ then define

$$(\tau^* c)_{U_{\alpha_0},\ldots,U_{\alpha_k}} = C_{\tau U_{\alpha_0},\ldots,\tau U_{\alpha_k}}$$

Then $\tau^* \circ \delta = \delta \circ \tau^*$. This implies that τ^* induces a map

$$\tau^*\colon H^k(\mathcal{V};G) \to H^k(\mathcal{U};G) .$$

τ^* is independent of the choice of τ (c.f. Hirzburch [10]). Set $\tau^{\mathcal{U}}_{\mathcal{V}}$ equal to the map

$$\tau^*\colon H^k(\mathcal{V};G) \to H^k(\mathcal{U};G) .$$

Then, if $\mathcal{V} < \mathcal{W} < \mathcal{U}$,

$$\tau_V^W \circ \tau_W^U = \tau_V^U$$

by the above observations. We define

$$H^k(M;G) = \varinjlim H^k(\mathcal{U};G) \quad .$$

That is, form the disjoint union of the $H^k(\mathcal{U};G)$. If $a \in H^k(\mathcal{U};G)$ and $b \in H^k(\mathcal{V};G)$ then we say:

$$a \sim b$$

if there is

$$\mathcal{W} > \mathcal{U} \quad \text{and} \quad \mathcal{V}$$

such that

$$\tau_{\mathcal{U}}^{\mathcal{W}} a = \tau_{\mathcal{V}}^{\mathcal{W}} b \quad .$$

Then $a \sim b$ is an equivalence relation and $\varinjlim H^k(\mathcal{U};G)$ is the set of all equivalence classes. $H^k(M;G)$ is an abelian group under the obvious operations.

<u>Definition 1.1</u>. If $U \subset M$ is open, then we say that U is (C^∞) contractible if there is a C^∞ map

$$F: [0,1] \times U \to U$$

so that $F(0,x) = x$ for all $x \in U$ and $F(1,x) \equiv x_0$ for all $x \in U$. An open covering $\mathcal{U} = \{U_\alpha\}_{\alpha \in I}$ is said to be contractible if, whenever

$$\alpha_1, \ldots, \alpha_k \in I$$

and

$$U_{\alpha_1} \cap \cdots \cap U_{\alpha_k} \neq \emptyset ,$$

then

$$U_{\alpha_1} \cap \cdots \cap U_{\alpha_k}$$

is contractible.

Theorem 1.2. If $\mathcal{U}$ is an open covering of M then $\mathcal{U}$ has a contractible refinement.

This result is proved by putting a Riemannian structure on M and taking for each $U \in \mathcal{U}$, $p \in U$, $N_{P,U}$ a convex neighborhood of p in U.

$$\mathcal{V} = \{N_{P,U}\}_{\substack{U\in\mathcal{U} \\ P\in U}}$$

is a contractible refinement. See Helgason [8] for the existence (and definition of convex neighborhoods).

Theorem 1.3. a) If $\mathcal{U}$ is a contractible covering of M then $H^k(\mathcal{U};G)$ is isomorphic with $H^k(M;G)$.

b) If G is the additive group of $\mathbb{R}$ or $\mathbb{C}$ we denote $H^k(M;G)$ by $H^k(M;\mathbb{R})$ or $H^k(M;\mathbb{C})$.

$H^k(M;G)$ is canonically isomorphic with $H_d^k(M;G)$. We denote the isomorphism by

$$d\cdot R\colon H_d^k(M;K) \to H^k(M;K) \quad .$$

Of course, this theorem is well known. a) is due to A. Weil [29] and b) is the de Rham theorem. We give a proof of b) (and a) for $K = \mathbb{R}$ or $\mathbb{C}$) for $k = 0,1$ or 2 since we will be using explicit forms of the isomorphisms asserted to exist in b). We need a special case of Theorem 1.3.

Lemma 1.4 (Poincaré Lemma, c.f., F. Warner [27]). If $U \subset M$ is an open contractible set and $\omega \in \mathcal{D}^k$, $d\omega = 0$, then there is $\eta \in \mathcal{D}^{k-1}$ so that $\omega = d\eta$.

We now begin the proof of b) for $k = 0,1$ or 2. We fix an arbitrary contractible covering

$$\{U_\alpha\}_{\alpha\varepsilon I} = \mathcal{U}$$

of M. We note that if we have shown

$$H^k(\mathcal{U};K) \equiv H_d^k(M;K)$$

then we will have shown both a) and b).

$k = 0$: If $c \in C^0(\mathcal{U};K)$, $\delta c = 0$, then

$$\delta c_{\alpha_0,\alpha_1} = c_{\alpha_0} - c_{\alpha_1} = 0$$

on $U_{\alpha_0} \cap U_{\alpha_1}$. Thus if

$$U_{\alpha_0} \cap U_{\alpha_1} \neq \emptyset ,$$

then

$$c_{\alpha_0} = c_{\alpha_1} .$$

Hence, c defines a constant on M. Hence, $H^0(\mathcal{U};K) \equiv K$. If $f \in \mathcal{D}^0$ and $df = 0$, then f is constant (M is connected). Hence,

$$H^0_d(M;K) \equiv K .$$

$k = 1$: Let $\omega \in \mathcal{D}^1$, $d\omega = 0$. Since U_α is contractible for each $\alpha \in I$ we have, by Lemma 1.4, $f_\alpha \in C^\infty(U_\alpha;K)$ so that

$$df_\alpha = \omega|_{U_\alpha}$$

for $\alpha \in I$.

If

$$U_{\alpha_0} \cap U_{\alpha_1} \neq \emptyset ,$$

then

$$d(f_{\alpha_0} - f_{\alpha_1}) = 0$$

on $U_{\alpha_0} \cap U_{\alpha_1}$.

Hence,

$$\left(f_{\alpha_0} - f_{\alpha_1}\right)\Big|_{U_{\alpha_0} \cap U_{\alpha_1}} = c_{\alpha_0,\alpha_1} \in K \quad .$$

Clearly, $c \in C^1(\mathcal{U};K)$ and $\delta c = 0$. If for each $\alpha \in I$, $g_\alpha \in C^\infty(U_\alpha;K)$ has been chosen so that

$$dg_\alpha = \omega|_{U_\alpha}$$

then $d(g_\alpha - f_\alpha) = 0$. Hence,

$$(g_\alpha - f_\alpha)|_{U_\alpha} = \lambda_\alpha \in K \quad .$$

If

$$b_{\alpha_0,\alpha_1} = (g_{\alpha_0} - g_{\alpha_1})\Big|_{U_{\alpha_0} \cap U_{\alpha_1}} \quad ,$$

then

$$b_{\alpha_0,\alpha_1} = c_{\alpha_0,\alpha_1} + \lambda_{\alpha_0} - \lambda_{\alpha_1} = c + \delta\lambda \quad .$$

Hence, the class of $c, [c]$ in $H^1(\mathcal{U};K)$ depends only on $\omega \in \mathcal{D}^1$, $d\omega = 0$. Clearly, if $\omega = df$ then $[c] = 0$. We therefore have a map

$$H^1_d(M;K) \to H^1(\mathcal{U};K) \quad .$$

If $d\omega = 0$, $\omega \in \mathcal{D}^1$ and $\omega \mapsto c$ with $[c] = 0$, then

$$c_{\alpha_0,\alpha_1} = b_{\alpha_0} - b_{\alpha_1} .$$

Set $\tilde{f}_\alpha = f_\alpha - b_\alpha$. Then

$$\left(\tilde{f}_{\alpha_0} - \tilde{f}_{\alpha_1}\right)\Big|_{U_{\alpha_0} \cap U_{\alpha_1}} = 0 .$$

Hence, if we set

$$\tilde{f}(x) = \tilde{f}_\alpha(x)$$

$$\text{for} \quad x \in U_\alpha ,$$

$\tilde{f} \in C^\infty(M;K)$ and $d\tilde{f} = \omega$. Hence $[\omega] \in H^1_d(M;K) = 0$. This implies that the map

$$H^1_d(M;K) \to H^1(\mathcal{U};K)$$

defined above is injective. To complete the proof of b) in the case $k = 1$ we must show that the map is surjective. Let $c \in C^1(\mathcal{U};K)$ and $\delta c = 0$. Let $\{\phi_\alpha\}_{\alpha \in I}$ be a partition of unity subordinate to $\mathcal{U}$ (c.f., F. Warner [27]). Set

$$f_\alpha = \sum_{\substack{\gamma \in I \\ U_\alpha \cap U_\gamma \neq \emptyset}} \phi_\alpha c_{\alpha,\gamma} .$$

Since $\delta c = 0$ it is easily checked that

$$(f_\alpha - f_\beta)\Big|_{U_\alpha \cap U_\beta} \equiv c_{\alpha\beta} .$$

Set

$$\omega_\alpha = df_\alpha$$

for $\alpha \varepsilon I$,

Then

$$\omega_\alpha \varepsilon \mathcal{D}^1(U_\alpha;K) .$$

$$(\omega_\alpha - \omega_\beta)\big|_{U_\alpha \cap U_\beta} = df_\alpha - df_\beta$$

$$= d(f_\alpha - f_\beta) = 0 .$$

Thus there is $\omega \varepsilon \mathcal{D}^1$ so that

$$\omega|_{U_\alpha} = \omega_\alpha$$

Now, following the above argument, we see that $[\omega] \to [c]$. This completes the proof in the case $k = 1$.

$k = 2$: If $\omega \varepsilon \mathcal{D}^2$ and $d\omega = 0$ then there exists η_α so that

$$\omega|_{U_\alpha} = d\eta_\alpha$$

$$\eta_\alpha \varepsilon \mathcal{D}^1(U_\alpha;K) \qquad (U_\alpha \text{ is contractible}) .$$

If $U_\alpha \cap U_\beta \neq \emptyset$, then

$$d(\eta_\alpha - \eta_\beta)\big|_{U_\alpha \cap U_\beta} = \omega|_{U_\alpha} - \omega|_{U_\beta} = 0 .$$

Hence,

$$\eta_\alpha - \eta_\beta = df_{\alpha\beta} \quad ,$$

$$f_{\alpha\beta} \in C^\infty(U_\alpha \cap U_\beta; K) \quad .$$

If $U_\alpha \cap U_\beta \cap U_\gamma \neq \emptyset$, then

$$d(f_{\beta\gamma} - f_{\alpha\gamma} + f_{\alpha\beta}) = \eta_\beta - \eta_\gamma - \eta_\alpha + \eta_\gamma + \eta_\alpha - \eta_\beta$$

$$= 0 \quad .$$

Hence,

$$(f_{\beta\gamma} - f_{\alpha\gamma} + f_{\alpha\beta})\Big|_{U_\alpha \cap U_\beta \cap U_\gamma} \equiv C_{\alpha,\beta,\gamma} \in K \quad .$$

Clearly, c defines an element in $C^2(\mathcal{U};K)$ and $\delta c = 0$. Arguing as in the case $k = 1$, we see that the map $\omega \to [c]$ is well defined and induces a map

$$H_d^2(M;K) \to H^2(\mathcal{U};K)$$

To prove the injectivity we note that if $c = \delta u$ and defining

$$f_{\alpha\beta} = f_{\alpha\beta} - u_{\alpha\beta}$$

then

$$\tilde{f}_{\beta\gamma} - \tilde{f}_{\alpha\gamma} + \tilde{f}_{\alpha\beta} \equiv 0$$

on $U_\alpha \cap U_\beta \cap U_\gamma$. Set

$$f_\alpha = \sum_{\substack{\gamma \varepsilon I \\ U_\alpha \cap U_\gamma \neq \emptyset}} \phi_\gamma f_{\alpha\gamma}$$

where $\{\phi_\alpha\}_{\alpha\varepsilon I}$ is a partition of unity subordinate to $\mathcal{U}$. Then it is easy to see that

$$(f_\alpha - f_\beta)\Big|_{U_\alpha \cap U_\beta} = \tilde{f}_{\alpha,\beta} .$$

Define

$$\tilde{\eta}_\alpha = \eta_\alpha - df_\alpha .$$

Then

$$\tilde{\eta}_\alpha - \tilde{\eta}_\beta = 0$$

on $U_\alpha \cap U_\beta$. Hence, there is $\tilde{\eta} \varepsilon \mathcal{D}^1$ so that

$$\tilde{\eta}|_{U_\alpha} = \tilde{\eta}_\alpha .$$

Clearly,

$$d\tilde{\eta} = \omega.$$

Hence, the map $H^1_d(M;K) \to H^1(\mathcal{U};K)$ is injective.

Set $c \varepsilon C^2(\mathcal{U};K)$, $\delta c = 0$. We show that there is $\omega \varepsilon \mathcal{D}^2$ so that

$$\omega|_{U_\alpha} = d\eta_\alpha ,$$

$$(\eta_\alpha - \eta_\beta)\Big|_{U_\alpha \cap U_\beta} = df_{\alpha\beta}$$

and

$$f_{\alpha\beta} - f_{\alpha\gamma} + f_{\alpha\beta} = c_{\alpha\beta\gamma}$$

on $U_\alpha \cap U_\beta \cap U_\gamma$. Define $f_{\alpha\beta}$ on $U_\alpha \cap U_\beta$ by

$$f_{\alpha\beta} = \sum_{\gamma\varepsilon I} \left(\phi_\gamma \Big|_{U_\alpha \cap U_\beta \cap U_\gamma}\right) c_{\alpha\beta\gamma}$$

Define

$$\eta_\alpha = \sum_{\gamma\varepsilon I} \phi_\gamma \, df_{\alpha\gamma}\Big|_{U_\alpha \cap U_\gamma}$$

Set

$$\omega_\alpha = d\eta_\alpha \quad .$$

Then

$$(\omega_\alpha - \omega_\beta)\Big|_{U_\alpha \cap U_\beta} = 0 \quad .$$

Hence there is $\omega \varepsilon \mathcal{D}^2$ so that

$$\omega\Big|_{U_\alpha} = \omega_\alpha$$

Clearly, $\omega \to [c]$. This completes the proof in the case $k = 2$.

Of course, the above arguments will work for any k. But they become progressively more complicated as k gets larger. See A. Weil [29] for a proof along these lines for any k.

2. LINE BUNDLES

Definition 2.1. Let M be a C^∞ manifold. Then a line bundle over M is a pair (L,p) consisting of a C^∞ manifold, L, and a surjective C^∞ mapping $p: L \to M$ such that

1) $L_x = p^{-1}(x)$ is a complex one dimensional vector space

2) If $x \in M$ there is a neighborhood U of x in M and a diffeomorphism in $\Phi: p^{-1}(U) \to U \times \mathbb{C}$ so that if $\ell \in p^{-1}(y)$,

$$\Phi(\ell) = (y, \phi_y(\ell))$$

and

$$\phi_y: L_y \to \mathbb{C}$$

is complex linear.

Note. If in 1) we said that $L_x = p^{-1}(x)$ is an n-dimensional complex vector space, 1), 2) would define a vector bundle over M. Many of the results we will develop for line bundles have analogues for vector bundles. Since we will really have no need for general vector bundles in these notes, we study only the case of line bundles.

Lemma 2.2. If U is contractible and (L,p) is a line bundle over M, then there exists $\Phi: p^{-1}(U) \to U \times \mathbb{C}$ satisfying 2) of Definition 2.1.

For a proof of this fact, see, for example, 1.4.8, p. 10 of Wallach [26].

Lemma 2.2 implies that if (L,p) is a line bundle over M and if

$$\mathcal{U} = \{U_\alpha\}_{\alpha \varepsilon I}$$

is a contractible covering of M then there is, for each $\alpha \varepsilon I$,

$$\Phi_\alpha : p^{-1}(U_\alpha) \to U_\alpha \times \mathbb{C}$$

so that 2) of Definition 2.1 is satisfied.

If $U_\alpha \quad U_\beta \neq \emptyset$, then

$$\Phi_\alpha \circ \Phi_\beta^{-1} : U_\alpha \cap U_\beta \times \mathbb{C} \to U_\alpha \cap U_\beta \times \mathbb{C} \quad .$$

By 2) above we have

1) $$(\Phi_\alpha \circ \Phi_\beta^{-1})(x,v) = (x, g_{\alpha\beta}(x)v)$$

$$\text{for} \quad x \varepsilon U_\alpha \cap U_\beta \quad \text{and} \quad v \varepsilon \mathbb{C} \quad .$$

Furthermore,

$$g_{\alpha\beta} : U_\alpha \cap U_\beta \to \mathbb{C}^x = \mathbb{C} - \{0\}$$

is C^∞.

It is easily checked that

2) $$g_{\alpha\beta} \cdot g_{\beta\gamma} = g_{\alpha\gamma}$$

on $U_\alpha \cap U_\beta \cap U_\gamma$.

<u>Note</u>. 2) is just the condition $\delta g = 0$ written multiplicatively.

Since $U_\alpha \cap U_\beta$ is contractible, there is

$$h_{\alpha\beta} \in C^\infty(U_\alpha \cap U_\beta ; \mathbb{C})$$

so that

3) $$g_{\alpha\beta} = \exp(2\pi i h_{\alpha\beta}) \quad .$$

Now 2) implies that

If $x \in U_\alpha \cap U_\beta \cap U_\gamma$ then

4) $$h_{\alpha\beta}(x) + h_{\beta\gamma}(x) - h_{\alpha\gamma}(x) \equiv c_{\alpha\beta\gamma} \in \mathbb{Z}$$
for $x \in U_\alpha \cap U_\beta \cap U_\gamma$.

Clearly, $c \in C^2(\mathcal{U};\mathbb{Z})$ and $\delta c = 0$.

<u>Lemma 2.3</u>. Let $K(L)$ denote the class of c in $H^2(M;\mathbb{Z})$. Then $K(L)$ is independent of the choices made in its definition. $K(L)$ is called the characteristic (or Chern) class of L.

Definition 2.4. Let (L_i, p_i), $i = 1,2$ be line bundles over M. Then a homomorphism $h: L_1 \to L_2$ is a C^∞ map h of L_1 to L_2 so that

$$h((L_1)_x) = C(L_2)_x \quad ,$$

$$x \in M$$

and

$$h: (L_1)_x \to (L_2)$$

is $\mathbb{C}$-linear. h is called an isomorphism if h is a surjective diffeomorphism. If there exists an isomorphism between L_1 and L_2 we say L_1 and L_2 are isomorphic.

Example 2.5. Let $M \times \mathbb{C}$ be made into a line bundle by defining

$$p(x,v) = x$$

and

$$\lambda \cdot (x,v) = (x,\lambda v) \quad ,$$

$$\lambda \in \mathbb{C} \quad .$$

We call any line bundle isomorphic with $M \times \mathbb{C}$ trivial.

Theorem 2.6. (L,p) is trivial if and only if $K(L) = 0$.

<u>Proof</u>. If (L,p) is trivial, then we have

$$\phi: L \to M \times \mathbb{C}$$

a vector bundle isomorphism. Taking

$$\Phi_\alpha = \phi\big|_{p^{-1}(U_\alpha)}$$

we have $g_{\alpha\beta} \equiv 1$, hence, $c = 0$. Thus $K(L) = 0$.

Suppose that $K(L) = 0$. Then

$$c_{\alpha\beta\gamma} = d_{\beta\gamma} - d_{\alpha\gamma} + d_{\alpha\beta}$$

$$\text{with} \quad d \in C^1(\mathcal{U};\mathbb{Z}) \quad .$$

Set

$$\tilde{h}_{\alpha\beta} = h_{\alpha\beta} - d_{\alpha\beta}$$

(we are using the notation in the derivation of 1), 2), 3), 4) above). Then

$$\tilde{h}_{\alpha\beta} + \tilde{h}_{\beta\gamma} - \tilde{h}_{\alpha\gamma} = 0$$

on $U_\alpha \cap U_\beta \cap U_\gamma$. Let $\{\phi_\alpha\}_{\alpha\in I}$ be a partition of unity subordinate to $\mathcal{U}$. Set

$$f_\alpha = \sum_\gamma \phi_\gamma \tilde{h}_{\alpha\gamma} \quad .$$

Then $f_\alpha \in C^\infty(U_\alpha;\mathbb{C})$ and $\tilde{h}_{\alpha\beta} = f_\alpha - f_\beta$ on $U_\alpha \cap U_\beta$. Set

$$\lambda_\alpha(x) = \exp(2\pi i f_\alpha) \quad .$$

Then $g_{\alpha\beta} = \lambda_\alpha \lambda_\beta^{-1}$. Define

$$\Phi_\alpha : p^{-1}(U_\alpha) \to U_\alpha \times \mathbb{C}$$

by

$$\Phi_\alpha(v) = (I \times \lambda_\alpha^{-1}(x))\phi_\alpha(v)$$

$$\text{for} \quad v \in p^{-1}(x) \quad .$$

Then $\Phi_\alpha(v) = \Phi_\beta(v)$ if $v \in p^{-1}(U_\alpha \cap U_\beta)$. Thus, setting $\Phi(v) = \Phi_\alpha(v)$ if $v \in p^{-1}(U_\alpha)$, $\Phi: L \to M \times \mathbb{C}$ is a homomorphism. Φ is clearly injective and a local diffeomorphism. Hence Φ is an isomorphism. Q.E.D.

3. CONNECTIONS

In general, contractible coverings are very complicated objects. It is therefore reasonable to try to compute characteristic classes in another way.

We first note that if $c \in C^k(\mathcal{U};\mathbb{Z})$ and $\delta c = 0$, then $c \in C^k(\mathcal{U};K)$ ($k = \mathbb{R}$ or $\mathbb{C}$) and $\delta c = 0$. If $c = \delta b$, $b \in C^{k-1}(\mathcal{U};\mathbb{Z})$, then $b \in C^{k-1}(\mathcal{U};K)$. Hence, we have a map

$$i: H^k(M;\mathbb{Z}) \to H^k(M;K) \quad ,$$

$K = \mathbb{R}$ or $\mathbb{C}$.

Definition 3.1. Let $K = \mathbb{R}$ or $\mathbb{C}$. If $\xi \in i(H^2(M;\mathbb{Z}))$ then ξ is said to be integral. If $\omega \in \mathcal{D}^2(M;K)$ and d.R.$[\omega]$ is integral, we say that ω is integral.

Definition 3.2. Let (L,p) be a line bundle over M. Then a cross-section of L is a C^∞ map $f: M \to L$ so that

$$p(f(x)) = x ,$$

$$x \in M .$$

Let $C^\infty(M;L)$ denote the space of all cross-sections of L.

If $f \in C^\infty(M;L)$ and $\phi \in C^\infty(M;\mathbb{C})$, then

$$(\phi f)(x) = \phi(x)f(x)$$

defines a cross-section of L.

Definition 3.3. Let

$$L \xrightarrow{p} M$$

be a line bundle over M. Then a connection on L is a $\mathbb{C}$-bilinear pairing for each $x \in M$ of $T(M)_x$ (the tangent space of M at x) with $C^\infty(M;L)$ into L_x.

$$(v,f) \mapsto \nabla_v f \in L_x ,$$

$$v \in T(M)_x, \quad f \in C^\infty(M;L)$$

so that

i) $\nabla_v(\phi f) = (v \cdot \phi) f(x) + \phi(x) \nabla_v f$

for $\phi \in C^\infty(M;\mathbb{C})$, $f \in C^\infty(M;L)$, $v \in T(M)_x$

ii) If $X \in \mathcal{X}(M)$ (the vector fields on M) and $f \in C^\infty(M;L)$, then

$$\left(x \mapsto \nabla_{X_x} f \underset{\text{def}}{\equiv} (\nabla_X f)(x)\right) \in C^\infty(M;L)$$

Lemma 3.4. If

$$L \xrightarrow{p} M$$

then there exists a connection on L.

Proof. Let $\mathcal{U} = \{U_\alpha\}_{\alpha \in I}$ be an open covering on M so that there exists

$$\Phi_\alpha : p^{-1}(U_\alpha) \to U_\alpha \times \mathbb{C}$$

a line bundle isomorphic for each $\alpha \in I$. Define

$$S_\alpha(x) = \Phi_\alpha^{-1}(x,1)$$

for $x \in U$

If $f \in C^\infty(M;L)$ and $x \in U_\alpha$, then

$$f(x) = h_\alpha(x) S_\alpha(x) \quad ,$$

$$h_\alpha \in C^\infty(U_\alpha;\mathbb{C}) \quad .$$

Set

$$\nabla_v^\alpha f = (v \cdot h_\alpha(x)) S_\alpha(x)$$

for $v \in T(M)_x$, $x \in U_\alpha$

$$\nabla_v^\alpha f = 0$$

if $v \in T(M)_x$, $x \notin U_\alpha$

Let $\{\phi_a\}_{\alpha \varepsilon I}$ be a partition of unity subordinate to U. Define

$$\nabla_v f = \sum_{\alpha \varepsilon I} \phi_\alpha(x) \nabla^\alpha f$$

for $v \in T(M)$, $f \in C^\infty(M;L)$

It is easy to check that ∇ defines a connection on L. Q.E.D.

Now let $L \xrightarrow{p} M$ be a line bundle and let ∇ be a connection on L. Let $\{U_\alpha\}_{\alpha \varepsilon I}$ be an open covering of M so that there exists

$$\Phi_\alpha : p^{-1}(U_\alpha) \to U_\alpha \times \mathbb{C}$$

a line bundle isomorphism. Define

$$S_\alpha(x) = \Phi_\alpha^{-1}(x,1)$$

$x \in U_\alpha$

If $f \varepsilon C^{\infty}(M;L)$, then

$$f|_{U_\alpha} = \phi_\alpha S_\alpha \quad , \quad \phi_\alpha \varepsilon C^{\infty}(U_\alpha;\mathbb{C}) \quad .$$

Hence

$$\nabla_v f = (v\cdot\phi_\alpha)S_\alpha(x) + \phi_\alpha(x)\nabla_v S_\alpha$$

for $x \varepsilon U_\alpha$.

The connection is therefore specified by $\nabla_v S_\alpha$. But

$$S_\alpha(x) \neq 0$$

for $x \varepsilon U_\alpha$

Hence,

$$\nabla_v S_\alpha = \theta_\alpha(v)S_\alpha$$

for $v \varepsilon T(M)_x$, $x \varepsilon U_\alpha$

Clearly, $\theta_\alpha \varepsilon \mathcal{D}^1(U_\alpha;\mathbb{C})$. If $x \varepsilon U_\alpha \cap U_\beta$ then

$$S_\alpha(x) = g_{\alpha\beta}(x)S_\alpha(x) \quad .$$

Hence

$$\nabla_v S_\alpha = (v\cdot g_{\alpha\beta})S_\beta + g_{\alpha\beta}\nabla_v S_\beta$$

for $v \varepsilon T(M)_x$, $x \varepsilon U_\alpha \cap U_\beta$

We therefore see

I) $\theta_\alpha(v) = (v \cdot g_{\alpha\beta}) g_{\alpha\beta}^{-1} + \theta_\beta(v)$

for $v \in T(M)_x$, $x \in U_\alpha \cap U_\beta$.

That is,

II) $\theta_\alpha = g_{\alpha\beta}^{-1} dg_{\alpha\beta} + \theta_\beta$

on $U_\alpha \cap U_\beta$.

Now, $d(g_{\alpha\beta}^{-1} dg_{\alpha\beta}) = 0$ on $U_\alpha \cap U_\beta$. Hence, $d\theta_\alpha = d\theta_\beta$ on $U_\alpha \cap U_\beta$. This means that we can define $\omega \in \mathcal{D}^2(M;\mathbb{C})$ by

$$\omega|_{U_\alpha} = d\theta_\alpha$$

for $\alpha \in U_\alpha$.

<u>Lemma 3.5</u>. ω is independent of all choices used to define it.

$$\nabla_X \nabla_Y f - \nabla_Y \nabla_X f - \nabla_{[X,Y]} f = \omega(X,Y) f$$

for $X, Y \in \mathcal{X}(M)$ and $f \in C^\infty(M;L)$.

<u>Proof</u>. It is clearly enough to prove the formula in the statement of the lemma.

We first note that

$$\nabla_X \nabla_Y (\phi f) - \nabla_Y \nabla_X (\phi f) - \nabla_{[X,Y]} (\phi f) = \phi(\nabla_X \nabla_Y f - \nabla_Y \nabla_X f - \nabla_{[X,Y]} f)$$

for $\phi \in C^\infty(M;L)$.

Hence, it is enough to prove the formula for $f = S_\alpha$ on U_α.

$$\begin{aligned}\nabla_X\nabla_Y S_\alpha - \nabla_Y\nabla_X S_\alpha - \nabla_{[X,Y]}S_\alpha &= \nabla_X(\theta_\alpha(Y)S_\alpha) - \nabla_Y(\theta_\alpha(X)S_\sigma) \\ &\quad - \theta_\alpha([X,Y])S_\alpha \\ &= (X\cdot\theta_\alpha(Y))S_\alpha + \theta_\alpha(Y)\theta_\alpha(X)S_\alpha \\ &\quad - (Y\cdot\theta_\alpha(X))S_\alpha - \theta_\alpha(X)\theta_\alpha(Y)S_\alpha \\ &\quad - \theta_\alpha([X,Y])S_\alpha \\ &= (X\cdot\theta_\alpha(Y) - Y\cdot\theta_\alpha(X) - \theta_\alpha([X,Y]))S_\alpha\end{aligned}$$

The lemma now follows since if $\theta \in \mathcal{D}^1(M;K)$,

$$d\theta(X,Y) = X\theta(Y) - Y\theta(X) - \theta([X,Y])$$

(Here we use

$$(dx_i \wedge dx_j)\left(\frac{\partial}{\partial x_i}, \frac{\partial}{\partial x_j}\right) = 1$$

to fix our definition of $\wedge$.)

<u>Definition 3.6</u>. ω is called the curvature form of L,∇ and also denoted

$$\operatorname{curv}(L,\nabla)$$

Clearly, $d(\operatorname{curv}(L,\nabla)) = 0$.

<u>Lemma 3.7</u>. $\frac{1}{2\pi i}$ d.R. $(\operatorname{curv}(L,\nabla)) = i(\mathcal{K}(L))$.

Proof. By the discussion above, if $\{U_\alpha\}_{\alpha\varepsilon I}$ is a contractible covering of M, then

$$\omega|_{U_\alpha} = d\theta_\alpha$$

$$(\theta_\alpha - \theta_\beta)\Big|_{U_\alpha \cap U_\beta} = dg_{\alpha\beta}/g_{\alpha\beta}$$

$$= d(\log g_{\alpha\beta}) \quad .$$

Thus,

$$\frac{1}{2\pi i}(\log g_{\alpha\beta} - \log g_{\alpha\gamma} + \log g_{\beta\gamma}) \equiv C_{\alpha\beta\gamma}$$

on $U_\alpha \cap U_\beta \cap U_\gamma$. $[c] \varepsilon H^2(M;\mathbb{C})$ is clearly $i(K(L))$. Q.E.D.

Definition 3.8. A Hermitian structure on L is a correspondence $x \mapsto H_x$ where H_x is a positive definite inner product on L_x such that if $f \varepsilon C^\infty(M;L)$, then

$$(x \mapsto H_x(f(x), f(x))) \varepsilon C^\infty(M) \quad .$$

A pair (L,H) consisting of a line bundle and a Hermitian structure on L is called a Hermitian line bundle. The same partition of unity argument used above to define a connection proves the existence of a Hermitian structure on L.

Definition 3.9. If L is a line bundle with connection ∇, and H is a Hermitian structure on L, then H is said to be ∇ invariant if

$$v\cdot H(f,g) = H(\nabla_v f,g) + H(f,\nabla_v g)$$

for $v \in T(M)$, $f,g \in C^\infty(M;L)$.

Lemma 3.10. Let L be a line bundle with connection ∇.

1) There is at most one (up to scalar multiple) ∇-invariant Hermitian structure on L.
2) If there exists a ∇-invariant Hermitian structure on L then $\frac{1}{2\pi i}\,\mathrm{curv}(L,\nabla)$ is a real two-form.
3) If $H^1(M,\mathbb{R}) = 0$ then there is a ∇-invariant Hermitian structure on L if and only if $\frac{1}{2\pi i}\,\mathrm{curv}(L,\nabla)$ is a real two-form.

Proof. Suppose that such an H exists. Let

$$\{U_\alpha\}_{\alpha\in I}\ ,\qquad \{S_\alpha\}_{\alpha\in I}$$

be as in the discussion leading to I) and II) above. Let

$$H(S_\alpha,S_\alpha) = \psi_\alpha$$

for $\alpha \in I$.

Set:

$$t_\alpha(x) = \psi_\alpha(x)^{-1/2}\, S_\alpha(x) \quad .$$

Then $H(t_\alpha, t_\alpha) \equiv 1$ on U_α. Let

$$\theta_\alpha(v)t_\alpha(x) = \nabla_v t_\alpha$$

for $v \in T(M)_x$, $x \in U_\alpha$

Now $0 = v$.

$$H(t_\alpha, t_\alpha) = H(\nabla_v t_\alpha, t_\alpha) + H(t_\alpha, \nabla_v t_\alpha)$$

$$= (\theta_\alpha(v) + \overline{\theta_\alpha(v)})$$

for $v \in T(M)_x$, $x \in U_\alpha$.

Hence, $\overline{\theta_\alpha} = -\theta_\alpha$. This says $\theta_\alpha = i\eta_\alpha$, $\eta_\alpha \in \mathcal{D}^1(U_\alpha)$. Hence $d\theta_\alpha = id\eta_\alpha$. Since

$$\omega|_{U_\alpha} = d\theta_\alpha$$

we see

$$\frac{1}{2\pi i} \cdot \mathrm{curv}(L, \nabla) \in \mathcal{D}^2(M) .$$

This proves (1).

Suppose that H' is another Hermitian structure on L so that ∇ is H'-invariant. Set

$$\xi_\alpha(x) = H'(t_\alpha, t_\alpha) .$$

If $v \in T(M)_x$, $x \in U_\alpha$, then

$$v \cdot \xi_\alpha = (\theta_\alpha(v) + \overline{\theta_\alpha(v)})\xi_\alpha = 0$$

Thus, $d\xi_\alpha = 0$ on U_α. Hence, ξ_α is a constant on U_α. This says

$$H'|_{U_\alpha} = c_\alpha H \ ,$$

$$c_\alpha \in \mathbb{R}, \quad c_\alpha > 0 \ .$$

The C^∞ condition clearly implies $c_\alpha = c_\beta$ for all α, β, proving 2).

Suppose finally that ∇ is a connection on L so that $\frac{1}{2\pi i}\,\mathrm{curv}(L,\nabla)$ is in $\mathcal{D}^2(M)$. Let $\{U_\alpha\}_{\alpha\in I}$, $\{S_\alpha\}_{\alpha\in I}$, $\{\theta_\alpha\}_{\alpha\in I}$ be as usual. $d\theta_\alpha = \mathrm{curv}(L,\nabla)$, thus $d(\theta_\alpha + \overline{\theta_\alpha}) = 0$. Hence

$$\theta_\alpha + \overline{\theta_\alpha} = df_\alpha$$

$$f_\alpha \in C^\infty(U_\alpha) \qquad (\theta_\alpha + \overline{\theta_\alpha} \in \mathcal{D}^1(M)) \ .$$

Set

$$\psi_\alpha = e^{f_\alpha} \ .$$

Define

$$H_x^\alpha(zS_\alpha(x),\, wS_\alpha(x)) = \psi_\alpha(x)z\overline{w} \ .$$

We assert that ∇ is H^α-invariant. Indeed,

$$\begin{aligned} v\cdot H\,(S_\alpha,S_\alpha) &= v\cdot\psi_\alpha = v\cdot e^{f_\alpha} \\ &= (v\cdot f_\alpha)e^{f_\alpha} = (\theta_\alpha(v) + \overline{\theta_\alpha(v)})\psi_\alpha \\ &= H^\alpha(\nabla_v S_\alpha, S_\alpha) + H^\alpha(S_\alpha, \nabla_v S_\alpha) \ . \end{aligned}$$

This implies that on $U_\alpha \cap U_\beta$,

$$H^\alpha = c_{\alpha\beta} H^\beta ,$$

$$c_{\alpha\beta} \in \mathbb{R}, \quad c_{\alpha\beta} > 0 .$$

Now, $c_{\alpha\beta} c_{\beta\gamma} = c_{\alpha\gamma}$. Thus, if

$$u_{\alpha\beta} = \log c_{\alpha\beta}$$

$$u \in C^1(\mathcal{U}, \mathbb{R}), \quad \delta u = 0 .$$

If $H^1(M;\mathbb{R}) = 0$ then $u_{\alpha\beta} = d_\alpha - d_\beta$, $d \in C^0(\mathcal{U}, \mathbb{R})$. Set

$$G^\alpha = e^{id_\alpha} H^\alpha .$$

Then

$$G^\alpha = e^{-d_\alpha} H^\alpha = e^{-d_\alpha} c_{\alpha\beta} H^\beta$$

$$= e^{(d_\beta - d_\alpha)} c_{\alpha\beta} G^\beta = G^\beta$$

on $U_\alpha \cap U_\beta$. Hence G^α, $\alpha \in I$ defines a ∇-invariant Hermitian structure on L. Q.E.D.

Lemma 3.11. Suppose that η is a real two-form on M so that $d\eta = 0$ and $d.R.([\eta]) \in i(H^2(M;\mathbb{Z}))$. Then there exists a line bundle L over M, a connection ∇ on L so that

$$\frac{1}{2\pi i} \operatorname{curv}(L,\nabla) = \eta .$$

Furthermore, if $H^1(M,\mathbb{R}) = 0$ there is a ∇-invariant Hermitian structure on L.

<u>Proof</u>. Let $\mathcal{U} = \{U_\alpha\}_{\alpha\varepsilon I}$ be a contractible covering of M. Let $\{\phi_\alpha\}_{\alpha\varepsilon I}$ be a partition of unity subordinate to $\mathcal{U}$.

$$\eta|_{U_\alpha} = d\theta_\alpha ,$$

$$\theta_\alpha \ \varepsilon\ \mathcal{D}^1(U_\alpha;\mathbb{R})$$

since $d\eta = 0$ on $U_\alpha \cap U_\beta$, $d(\theta_\alpha - \theta_\beta) = 0$. Thus,

$$(\theta_\alpha - \theta_\beta)\Big|_{U_\alpha \cap U_\beta} = dh_{\alpha\beta},$$

$$h_{\alpha\beta}\ \varepsilon\ C^\infty(U_\alpha \cap U_\beta;\ \mathbb{R}) \quad .$$

Set

$$g_{\alpha\beta} = \exp(2\pi i h_{\alpha\beta}) \quad .$$

Then since $d.R.[\eta]\ \varepsilon\ i(H^2(M,\mathbb{Z}))$ we see that we may assume

$$(h_{\beta\gamma} - h_{\alpha\gamma} + h_{\alpha\beta})\Big|_{U_\alpha \cap U_\beta \cap U_\gamma} \equiv C_{\alpha\beta\gamma}\ \varepsilon\ \mathbb{Z} \quad .$$

Hence,

$$\begin{aligned} g_{\alpha\gamma}g_{\gamma\beta} &= \exp(2\pi i(h_{\alpha\gamma} + h_{\gamma\beta}) \\ &= \exp(2\pi i(h_{\alpha\beta})) \\ &= g_{\alpha\beta} \quad . \end{aligned}$$

We construct L as follows: Let

$$\tau = \{(x,v,\alpha)|\ x\varepsilon U_\alpha,\ v\varepsilon\mathbb{C},\ \alpha\varepsilon I\} \subset M \times \mathbb{C} \times I$$

with the subspace topology (here I is given the discrete topology). Define

$$(x,v,\alpha) \equiv (y,w,\beta)$$

if and only if

$$x = y, \quad w = g_{\beta\alpha}(x)v \quad .$$

Let $L = T/\equiv$ with the quotient topology. Let

$$\tilde{p}(x,v,\alpha) = x \quad .$$

Then $\tilde{p}$ induces a continuous surjective map, $p: L \to M$. Let $[x,v,\alpha]$ denote the equivalence class of (x,v,α).

Define

$$c_1[x,v,\alpha] + c_2[x,w,\alpha] = [x,c_1v + c_2w,\alpha] \quad .$$

This is well defined and gives $L_x = p^{-1}(x)$ the structure of a one-dimensional vector space over $\mathbb{C}$.

Set

$$\Phi_\alpha([x,v,\alpha]) = (x,v) \quad .$$

Give L the C^∞ structure that makes

$$\Phi_\alpha: p^{-1}(U_\alpha) \to U_\alpha \times \mathbb{C}$$

a diffeomorphism. Let $S_\alpha(x) = [x,1,\alpha]$.

Define a connection on L as follows:

$$\nabla_v S_\alpha = 2\pi i \theta_\alpha(v) S_\alpha(x) \quad ,$$

$$x \in U \ , \quad v \in T(M)_x$$

On $U_\alpha \quad U_\beta$ we see $\theta_\alpha - \theta_\beta = d \log g_{\alpha\beta}$. Thus ∇_v defines a connection on L. By definition of ∇_v we have

$$d\theta_\alpha = 2\pi i \eta|_{U_\alpha} \quad .$$

Hence $2\pi i\eta = \text{curv}(L,\nabla)$. Now the last part of the lemma follows from Lemma 2.17 (3). Q.E.D.

Let $L^\times = \{\ell \neq 0 | \ \ell \in L\}$.

<u>Lemma 3.12</u>. Let (L,∇) be a line bundle with connection over M. Then there exists a unique $\theta \in \mathcal{D}^1(L^\times;\mathbb{C})$ so that if $U \subset M$ is open and $S \in C^\infty(U;L|_U)$, $S(x) \neq 0$ for $x \in U$, then

$$\nabla_v S = (S^*\theta)(v) S(x) \qquad \text{(S* the codifferential of S)}$$

$$\text{for} \quad x \in U, \quad v \in T(M)_x \quad .$$

θ is called the connection form of ∇.

Our proof of this lemma takes some preparation. There are easier proofs but we will need many of the concepts in this proof later. We note that the concepts developed relate the two standard definitions of a connection (c.f., Kobayashi-Nomizu [15]).

Definition 3.13. If $\sigma: [a,b] \to M$ is a C^∞ curve, then a cross-section of L over σ is a C^∞ map $S: [a,b] \to L$ so that $S(t) \in L_{\sigma(t)}$.

Lemma 3.14. Let $C^\infty(\sigma,L)$ denote the space of cross-sections of L over σ. Then there exists a unique map

$$\frac{D}{dt} : C^\infty(\sigma;L) \to C^\infty(\sigma;L)$$

satisfying

(1)
$$\frac{D}{dt}(fs) = f's + f\frac{Ds}{dt}$$
for $f \in C^\infty([a,b];\mathbb{C})$, $s \in C^\infty(\sigma;L)$

If $s \in C^\infty(M;L)$ and $\phi(t) = S(\sigma(t))$ then

(2)
$$\frac{D}{dt}\phi = \nabla_{\sigma(t)}S$$
$$\left(\sigma(t)f = \frac{d}{dt} f(\sigma(t))\Big|_{t=0}\right).$$

Proof. The proof is exactly the same as the proof of the existence of covariant differentiation along a curve in a Riemannian manifold (c.f., Milnor [19]).

Definition 3.15. $S \in C^\infty(\sigma;L)$ is said to be parallel if $DS/dt = 0$.

The existence and uniqueness theorem for ordinary differential equations implies that if $S_0 \in L_{\sigma(a)}$ then there exists a unique $S \in C^\infty(\sigma;L)$ so that $S(a) = S_0$ and $DS/dt = 0$.

If $S_0 \in L^\times$ and $\sigma: (-\epsilon,\epsilon)$ M is a C^∞ curve so that $p(S_0) = \sigma(0)$, let S be the parallel cross-section over σ so that $S(0) = S_0$. Let $H_{S_0} = \{\dot{S}(0) \mid S \text{ as above}\}$. The existence and uniqueness theorem for ordinary differential equations depending on parameters implies that $S_0 \to H_{S_0}$ defines a C^∞ sub-bundle of $T(L^\times)$. Furthermore,

$$\dim H_{S_0} = \dim M$$

for $S_0 \in L^\times$.

Let

$$V_{S_0} = \ker P_{*S_0} .$$

Then clearly,

$$T(L^\times)_{S_0} = H_{S_0} \oplus V_{S_0} .$$

Now $\mathbb{C}^\times = \mathbb{C} - \{0\}$ acts on $L^\times$ on the right by scalar multiplication:

$$v \cdot z = zv$$

for $v \in L^\times$, $z \in \mathbb{C}^\times$

Let

$$L^{\times} = \{\ell \varepsilon L \mid \ell \neq 0\} .$$

If $S_0 \varepsilon L^{x}$, define

$$\beta(S_0)\colon \mathbb{C}^{\times} \to L^{\times}$$

by

$$\beta(S_0)(z) = S_0 \cdot z .$$

We identify $T(\mathbb{C}^{\times})_1$ with $\mathbb{C}$ in the usual way.

Then

$$\beta(S_0)_{*1}\colon T(\mathbb{C}^{\times})_1 = \mathbb{C} \to T(L^{\times})_{S_0} .$$

Clearly,

$$\beta(S_0)_{*1}(\mathbb{C}) = V_{S_0} .$$

($\beta(S_0)_{*1}$ is the differential of $\beta(S_0)$ at 1.)

If $v \varepsilon T(L^{\times})_{S_0}$ then $v = v^H + v^V$ with $v^H \varepsilon H_{S_0}$, $v^V \varepsilon V_{S_0}$. The maps $v \mapsto v^H$, $v \mapsto v^V$ are C^{∞}.

Define

$$\theta(v) = (\beta(S_0)_{*1})^{-1}(v^V) .$$

Then $\theta \varepsilon \mathcal{D}^1(L^{\times},\mathbb{C})$. This is the θ of Lemma 3.12.

We must show that if $U \subset M$ is open and $S\colon U \to L^{\times}$ is a C^{∞} cross-section of $L|_U$, then

$$\nabla_v S = (S^*\theta)(v)S(x)$$

for $x \in U$, $v \in T(M)_x$.

Let $x \in U$, $v \in T(M)_x$. Let $\sigma: (-\epsilon, \epsilon) \to M$ be a C^∞ curve so that $\sigma(0) = x$, $\dot{\sigma}(0) = v$. Let ψ be the parallel cross-section along σ so that $\psi(0) = S(x)$. Then

$$S(\sigma(t)) = h(t)\psi(t) .$$

Set

$$\delta(t) = S(\sigma(t)) .$$

Then

$$\dot{\delta}(t) = \beta(\psi(t))_{*1}(h'(t)) + h(t)\dot{\psi}(t) .$$

Hence

$$\dot{\delta}(0) = \beta(S(0))_{*1}(h'(0)) + h(0)\dot{\psi}(t) .$$

This says

$$\dot{\delta}(0)^V = \beta(S(0))_{*1}(h'(0)) .$$

Hence $\theta(\dot{\delta}(0)) = h'(0)$. Now, $\dot{\delta}(0) = S_{*x}v$. Thus

$$(S^*\theta)(v) = h'(0) .$$

Hence

$$\begin{aligned} (S^*\theta)(v)S(x) &= h'(0)S(x) = h'(0)\psi(0) \\ &= h'(0)\psi(0) + h(0)\,\frac{D\psi}{dt}(0) \\ &= \nabla_{\dot{\sigma}(0)}S . \end{aligned}$$

Thus θ has the desired property.

The uniqueness of θ is a direct consequence of the fact that

$$T(L^{\times})_{S_0} = \{S_{*_x}v \mid v \in T(M)_x, \quad x \in U \text{ open}, \quad s \in C^{\infty}(U;L),\ S(y) \neq 0, \quad y \in U, \quad S(x) = S_0\} \quad .$$

Q.E.D.

Chapter 2

SYMPLECTIC GEOMETRY

In this chapter we begin with a short introduction to symplectic geometry: That is, Poisson brackets, the Darboux theorem, Hamiltonian vector fields. We then go on in Section 5 to the situation when the symplectic structure gives an integral cohomology class. We develop several results of Kostant relating the automorphisms of the symplectic structure and the automorphisms of the corresponding line bundle with connection. We then introduce (following Kostant) the notion of pre-quantization. These results put in elegant form earlier work of Von Hove [25]. Also a very similar discussion can be found in Souriau [24].

1. SYMPLECTIC MANIFOLDS

Definition 1.1. A symplectic manifold is a pair (M,ω) consisting of M a C^∞ manifold and $\omega \in \mathcal{D}^2(M;\mathbb{R})$, $d\omega = 0$ and such that ω_x is non-degenerate for each $x \in M$. (That is, if $x \in M$ and $\omega_x(v,z) = 0$ for all $z \in T(M)_x$, then $v = 0$.) ω is called a symplectic structure on M.

Examples. 1.2. If $\dim M = 2$ and M is orientable, then a symplectic structure on M is a volume form.

1.3. Let M be a manifold and let $T(M)^*$ be the cotangent bundle of M. Let $\pi\colon T(M)^* \to M$ be the projection. Define

$$\theta \in \mathcal{D}^1(T(M)^*;\mathbb{R})$$

as follows:

$$\theta(v) = \lambda(\pi_{*_\lambda}(v)) \quad .$$

We assert that $d\theta = \omega$ defines a symplectic structure on M. Indeed, let $U \subset M$ be open with local coordinates $x_1,\ldots,x_n$. Then we get local coordinates on $\pi^{-1}(U)$ by

$$\sum \lambda_i \, dx_{i_y} \to (x_1(y),\ldots,x_n(y),\ \lambda_1,\ldots,\lambda_n)$$

$$y \in U \quad .$$

$$\pi\left(\sum \lambda_i \, dx_{i_y}\right) = y \quad .$$

(a) $$\theta = \sum \lambda_i \, dx_i$$

$$\text{on } \pi^{-1}(U) \quad .$$

Indeed, in local coordinates

$$v = \sum a_i \frac{\partial}{\partial x_i} + \sum b_i \frac{\partial}{\partial x_i}$$

$$\pi_* \frac{\partial}{\partial x_i} = 0 \quad .$$

Hence,

$$\pi_{*\lambda}(v) = \sum a_i \frac{\partial}{\partial x_{i_x}}$$

This implies that

$$\theta(v) = \sum a_i \lambda_i = \sum \lambda_i \, dx_i(v) \quad .$$

Now,

$$d\theta = \sum d\lambda_i \wedge dx_i \quad ,$$

which is clearly non-degenerate.

The symplectic structure so defined on $T(M)^*$ is called the <u>canonical</u> <u>symplectic</u> structure.

1.4. Let G be a connected Lie group with Lie algebra $\mathfrak{g}$. Let $\mathfrak{g}^*$ be the space of all real linear forms on $\mathfrak{g}$. If $f \in \mathfrak{g}^*$, $g \in G$, define

$$g \cdot f = f \circ Ad(g)^{-1} \quad .$$

($Ad(g)$ is the differential of the inner automorphism $x \mapsto gxg^{-1}$.)

If $f \in \mathfrak{g}^*$, let

$$\theta_f = G \cdot f \quad .$$

Let

$$G_f = \{g \in G \mid g \cdot f = f\} \quad .$$

Then we make θ_f into a manifold by giving it the manifold structure of G/G_f. Now, θ_f is clearly a submanifold of $\mathcal{g}^*$. Hence we can identify $T(\theta_f)_x$ with a subspace of $\mathcal{g}^*$ for each $x \in \theta_f$.

Let $\mathcal{g}$ act on $\mathcal{g}^*$ by $X \cdot f = -f \circ \text{ad } X$ $(\text{ad } X \cdot Y = [X,Y])$. Then

$$T(\theta_f)_x = \mathcal{g} \cdot x = g \cdot \mathcal{g} \cdot f$$

$$\text{for } x = g \cdot f \quad .$$

If

$$v, z \in T(\theta_f)_x \quad ,$$

$$v = V \cdot x \ , \quad z = Z \cdot x$$

we define

$$\omega_x(v,z) = x([V,Z]) \quad .$$

We must show that ω_x is well defined. If $v = V' \cdot x$, $V' \in \mathcal{g}$ then $V' = V + V' - V$ and $(V'-V) \cdot x = 0$. Hence,

$$x([V',Z]) = x(V,Z]) + ((V-V') \cdot x)(Z)$$

$$= x([V,Z]) \quad .$$

By skew symmetry ω_x is well defined.

We next note that

$$\omega \in D^2(\theta_f, \mathbb{R}) \quad .$$

Indeed, if $X,Y \in g$, let $X^*,Y^* \in X(\theta_f)$ be defined by

$$X^*_y = X\cdot y \ , \qquad Y^*_y = Y\cdot y \ .$$

Then

$$\omega_y(X^*_y,Y^*_y) = y([X,Y])$$

which is clearly C^∞ in y.

Suppose that $v \in T(\theta_f)_x$ and $\omega_x(v,z) = 0$ for all $z \in T(\theta_f)_x$. Then, if

$$v = V\cdot x \ , \quad z = Z\cdot z \ ,$$

then

$$x\cdot([V,Z]) = 0$$

$$\text{for all } Z \in g \ .$$

Hence, $-V\cdot x = 0$. Hence $v = 0$.

Thus, to show that ω defines a symplectic structure on θ_f, we need only show that $d\omega = 0$. To see this we must show that if $X,Y,Z \in g$ then

$$d\omega(X^*,Y^*,Z^*) = 0 \ .$$

But

$$d\omega(X^*,Y^*,Z^*) = X^*\omega(Y^*,Z^*) - Y^*\omega(X^*,Z^*) + Z^*\omega(X^*,Y^*)$$
$$- \omega([X^*,Y^*],Z^*) + \omega([X^*,Z^*],Y^*) - \omega([Y^*,Z^*],X^*)$$

Now, if $\phi \in C^{\infty}(\theta_f)$, then

$$X^*_y = \frac{d}{dt}_{t=0} \phi(\exp tX \cdot y)$$

for $X \in g$, $y \in \theta_f$.

Hence

$$X^*_y \omega(Y^*, Z^*) = (X \cdot y)([Y,Z])$$

$$= -y([X,[Y,Z]]) \quad .$$

But then (since $[X^*,Y^*] = -[X,Y]^*$)

$$(d\omega)_y(X^*,Y^*,Z^*) = -y([X,[Y,Z]]) + y([Y,[X,Z]]) - y([Z,[X,Y]])$$

$$+ y([[X,Y],Z]) - y([[X,Z],Y]) + y([[Y,Z[,X])$$

$$= -2y([X,[Y,Z]]) + 2y([Y,[X,Z]]) - 2y([Z,[X,Y]])$$

But the Jacobi identity says

$$[X,[Y,Z]] = [[X,Y],Z] + [Y,[X,Z]] \quad .$$

Hence, $d\omega = 0$.

Hence (θ_f,ω) is a symplectic manifold. We denote ω by ω^{θ} or ω^{θ_f} or ω^f if we must deal with more than one orbit at a time.

We note that if $g \varepsilon G$, then $g^*\omega^f = \omega^f$. Thus, $g \varepsilon G$ acts on θ_f by diffeomorphisms preserving ω^f. We will see that under suitable conditions on G all homogeneous symplectic manifolds are (locally) of the form (θ_f, ω^f), $f \varepsilon \mathfrak{g}^*$.

We now give an example of this construction.

1.4.(a) $\mathfrak{g} = \mathfrak{su}(n+1) = \{X \varepsilon M_{n+1}(\mathbb{C}) \mid {}^t\overline{X} = -X,\ \mathrm{tr}\ X = 0\}$

$$G = SU(n+1) = \{g \varepsilon SL(n+1,\mathbb{C}) \mid {}^t\overline{g} = g^{-1}\}$$

Let

$$X = \begin{bmatrix} iI & 0 \\ \sigma & -in \end{bmatrix}$$

I the $n \times n$ identity matrix. Set

$$f(Y) = -\mathrm{tr}\ XY$$

for $Y \varepsilon \mathfrak{g}$.

If $g \varepsilon G$ and $g \cdot f = f$ then

$$(g \cdot f)(Y) = f(Y)$$

for all $Y \varepsilon \mathfrak{g}$

But then

$$\mathrm{tr}(g^{-1}YgX) = \mathrm{tr}\ YX$$

for all $Y \varepsilon \mathfrak{g}$.

Hence $gXg^{-1} = X$. Thus,

$$g = \left[\begin{array}{c|c} A & 0 \\ \hline 0 & b \end{array}\right]$$

$A \varepsilon U(n)$, $\det A = b^{-1}$.

$$\theta_f = SU(n+1)/U(n) = P^n(\mathbb{C}) .$$

1.4.(b) g,G as in 3(a). Take

$$X = \begin{bmatrix} iqI_p & 0 \\ 0 & -ipI_q \end{bmatrix} ,$$

$$p + q = n + 1$$

I_p the $p \times p$ identity matrix. Then

$$\theta_f = \frac{SU(n+1)}{S(U(p) \times U(q))}$$

$$S(U(p)\times U(q)) = \left\{ \left[\begin{array}{c|c} A & \\ \hline & B \end{array}\right] \;\middle|\; A \varepsilon U(p),\ B \varepsilon U(q),\ \det A\cdot\det B = 1 \right\}$$

If (M,ω) is a symplectic manifold, we can associate to each $f \varepsilon C^\infty(M)$ a vector field as follows:

(i) $$\omega_x(X_f, v) = v \cdot f$$

for $v \in T(M)_x$.

If $f, g \in C^\infty(M)$, define

$$\{f,g\} = \omega(X_f, X_g) = X_g \cdot f = -X_f \cdot g$$

Lemma 1.5.

$$\{f,\{g,h\}\} = \{\{f,g\},h\} + \{g,\{f,h\}\} \quad .$$

That is, $(C^\infty(M), \{\ ,\ \})$ is a Lie algebra. Furthermore,

$$X_{\{f,g\}} = [X_f, X_g] \quad .$$

Note. (i) defines X_f since ω_x is non-degenerate for each $x \in M$. Lemma 1.5 is the reason for the condition $d\omega = 0$.

Proof.

$$\{f,\{g,h\}\} = -X_f \omega(X_g, X_h) \quad .$$

Now,

$$\begin{aligned} 0 &= d\omega(X_f, X_g, X_h) \\ &= X_f \cdot \omega(X_g, X_h) - X_g \omega(X_f, X_h) + X_h \omega(X_f, X_g) - \omega([X_f, X_g], X_h) \\ &\quad + \omega([X_f, X_h], X_g) - \omega([X_g, X_h], X_f) \end{aligned}$$

$$= -\{f,\{g,h\}\} + \{g,\{f,h\}\} - \{h,\{f,g\}\}$$

$$+ [X_f,X_g]\cdot h - [X_f,X_h]\cdot g + [X_g,X_h]\cdot f$$

$$= -\{f,\{g,h\}\} + \{g,\{f,h\}\} - \{h,\{f,g\}\}$$

$$+\{f,\{g,h\}\} - \{g,\{f,h\}\} - \{f,\{h,g\}\}$$

$$+\{h,\{f,g\}\} + \{g,\{h,f\}\} - \{h,\{g,f\}\}$$

$$= \{f,\{g,h\}\} - \{g,\{f,h\}\} + \{h,\{f,g\}\}$$

Q.E.D.

Definition 1.6. $\{f,g\}$ is called the Poisson bracket of f and g.

2. THE DARBOUX THEOREM

In this section we develop results on the exterior derivative that we will use later in these notes and which are useful to give a particularly easy proof (due to Weinstein [32]) of Theorem 2.1 below.

Theorem 2.1 (Darboux). Let (M,ω) be a symplectic manifold. Let $p \in M$. Then there exists U an open neighborhood of p

in M and local coordinates $\{x_1,\ldots,x_k, y_1,\ldots,y_k\}$ on U so that

$$\omega|_U = \sum dx_i \wedge dy_i$$

In particular, $\dim M = 2k$. The coordinates $\{x_1,\ldots,x_k, y_1,\ldots,y_k\}$ are called canonical coordinates.

Note. Relative to a system of canonical coordinates

$$\{f,g\} = \sum \frac{\partial f}{\partial x_i}\frac{\partial g}{\partial y_i} - \sum \frac{\partial f}{\partial y_i}\frac{\partial g}{\partial x_i} \quad .$$

To prove this result it will be convenient to develop some facts about exterior derivatives, these facts will be useful later.

Let ω be a k-form; then we have already used

$$\begin{aligned} d\omega(X_1,\ldots,X_{k+1}) &= \sum_{i=1}^{k+1} (-1)^{i+1} X_i\omega(X_1,\ldots,\hat{X}_i,\ldots,X_{k+1}) \\ \text{(I)} \qquad &+ \sum_{i<j} (-1)^{i+1}\omega([X_i,X_j],X_1,\ldots,\hat{X}_i,\ldots,\hat{X}_j,\\ &\qquad \ldots,X_{k+1}) \quad . \end{aligned}$$

To prove (I) we note that the right hand side satisfies

$$\eta(f_1X_1,\ldots,f_{k+1}X_{k+1}) = f_1,\ldots,f_{k+1}\eta(X_1,\ldots,X_{k+1})$$

for $X_i \in \mathcal{X}(M)$, $f_i \in C^\infty(M)$.

This it is enough to check for $X_i = \partial/\partial x_i$ in a system of local coordinates.

If ω is a k-form and X is a vector field on M we define

$$(i(X)\omega)(X_1,\ldots,X_{k-1}) = \omega(X,X_1,\ldots,X_{k-1}) \quad .$$

Then clearly, $i(X)\omega \in \mathcal{D}^{k-1}$.

(II) $$d(i(X)\omega) = \mathcal{L}_X\omega - i(X)(d\omega)$$

where

$$(\mathcal{L}_X\omega)(X_1,\ldots,X_k) = X\omega(X_1,\ldots,X_k) - \sum_{i=1}^{k} \omega(X_1,\ldots,[X,X_i],\ldots,X_k)$$

This is an easy consequence of (I).

(III) Let X be a vector field on M and ω a k-form. Let ϕ_t be the local one-parameter group of diffeomorphisms generated by X. Then

$$\frac{d}{dt}(\phi_t^*\omega) = \phi_t^*(\mathcal{L}_X\omega) \quad .$$

This follows from

(IV) If $X,Y \in \mathcal{X}(M)$ and ϕ_t is the flow generated by X, then

$$\frac{d}{dt}\bigg|_{t=0} (\phi_{-t})_{*\phi_t(p)} Y_{\phi_t(p)} = [X,Y]_p \quad .$$

To prove (IV) we note

$$f(\phi_{-t}(q)) = f(q) - tX_q f + t^2 h(t,q)$$

by Taylor's theorem. Hence,

$$\phi_{-t_{*\phi_t(p)}} Y_{\phi_t(p)} f = Y_{\phi_t(p)} f \circ \phi_{-t}$$

$$= Y_{\phi_t(p)} f - tY_{\phi_t(p)}(Xf) + t^2 Y_{\phi_t(p)}(h(t,\cdot))$$

$$= Y_p f + tX_p(Yf) + t^2 u(t,p) - tY_p(Xf)$$

$$+ t^2 v(t,p) + t^2 Y_{\phi_t(p)}(h(t,\cdot)) \quad .$$

Taking $(d/dt)_{t=0}$ gives the result.

We need just a little more to prove the theorem. A time dependent vector field on M is a C^∞ map $X: (a,b) \times M \to T(M)$ so that $X(t,p) \in T(M)_p$.

The "flow" generated by a time dependent vector field is the solution (local) to

$$\frac{d}{dt} f(F_t(q)) = X(t,q)f \quad .$$

Time dependent vector fields can be handled in the same way as vector fields by the following trick: Define the vector field $\overline{X}$ on $(a,b) \times M$ by

$$\overline{X}(t,p) = \frac{d}{dt} + X(t,p)$$

(we look at $T((a,b) \times M)_{(s,p)}$ as $\mathbb{R}\frac{d}{dt} \oplus T(M)_p$.) Let $\overline{\phi}_s$ be the flow generated by $\overline{X}$ on $(a,b) \times M$. Then it is easy to check that

$$\overline{\phi}_s(t,q) = (t+s,\ H(s,t,q)) \quad .$$

Furthermore,

$$F_s(q) = H(s,0,q) \quad .$$

Thus in particular F_s is a local one-parameter family of diffeomorphisms. If $\omega \in \mathcal{D}^k(M)$ and $\pi\colon (a,b) \times M \to M$ is given by $\pi(t,q) = q$, then

$$\pi^* F_s^* \omega = \overline{\phi}_s^* \pi^* \omega \quad .$$

Hence,

$$\frac{d}{ds}\Big|_{s=0} \pi^* F_s^* \omega = \mathcal{L}_{\overline{X}} \pi^* \omega \quad .$$

We therefore see

$$\frac{d}{dt} F_t^* \omega = \mathcal{L}_{X_t} \omega \quad ,$$

$$X_t(q) = X(t,q) \quad .$$

(V) If X is a time dependent vector field and F_t is the flow generated by X, then if $\omega \in \mathcal{D}^k(M)$,

$$\frac{d}{dt} F_t^*\omega = \mathcal{L}_{X_t}\omega \quad .$$

We are now ready to give Weinstein's proof of Darboux's theorem.

We first note that since the theorem is local, we may assume $M = \mathbb{R}^n$, $p = 0$. Let

$$\omega = \sum_{i<j} a_{ij}\, dx_i \wedge dx_j$$

in standard coordinates. Let

$$\omega_1 = \sum_{i<j} a_{ij}(0)\, dx_i \wedge dx_j$$

Set

$$\tilde{\omega} = \omega_1 - \omega \quad , \qquad \omega_t = \omega + t\tilde{\omega}$$

for all t .

Now,

$$\omega_t = \sum_{i<j} (a_{ij} + t(a_{ij}(0) - a_{ij}))\, dx_i \wedge dx_j \quad .$$

Hence, $\omega_t(0) = \omega(0)$ which is non-degenerate. There is, therefore, an open ball V_1 around 0 so that $\omega_t(q)$ is non-degenerate for each $q \in V_1$, $0 \le t \le 1$.

Now, $d\tilde{\omega} = 0$. Since V_1 is contractible, there is $\alpha \in \mathcal{D}^1(V_1)$ so that

$$d\alpha = \tilde{\omega} .$$

We may assume $\alpha(0) = 0$.

Define $X_t \in \mathscr{X}(V_1)$ by

$$i(X_t)\omega_t = -\alpha .$$

Then X_t defines a time dependent vector field on V_1 for $-\epsilon < t < 1+\epsilon$ for some $t > 0$. By possibly shrinking V_1 (and observing that $(X_t(0) = 0$ for $0 \le t \le 1)$ we may assume $F_t(q)$ is defined for $0 \le t \le 1$, $q \in V_1$. Now,

$$\frac{d}{dt}(F_t^*\omega_t) = F_t^*\left(L_{X_t}\omega_t + \frac{d}{dt}\omega_t\right)$$

$$= F_t^*(d(i(X_t)\omega_t) + i(X_t)d\omega_t + \tilde{\omega})$$

$$= F_t^*(-d\alpha + \tilde{\omega}) = 0$$

Hence $F_t^*\omega_t$ is constant in t. This says

$$F_0^*\omega_0 = F_1^*\omega_1 .$$

Now $F_0^* = I$, $\omega_0 = \omega$. Hence

$$F_1^* \omega_1 = \omega .$$

This says that F_1 defines a chart so that relative to the coordinates defined by F_1, $z_1, \ldots, z_n$

$$= \sum_{i<j} b_{ij} \, dz_i \wedge dz_j$$

$$b_{ij} \in \mathbb{R}$$

To complete the proof, we need only show that if

$$B: \mathbb{R}^n \times \mathbb{R}^n \to \mathbb{R}$$

is a skew symmetric non-degenerate bilinear form, then

(a) $n = 2k$

(b) There is a basis $e_1, \ldots, e_{2k}$ of $\mathbb{R}^n$ so that

$$B(e_i, e_j) = 0$$

$$\text{if } i,j \leq k \quad \text{or} \quad i,j > k$$

and

$$B(e_i, e_{j+k}) = \delta_{ij}$$

$$\text{for } 1 \leq i,j \leq k .$$

This is proved easily by induction. Choose e_1 to be an arbitrary non-zero vector. Let e_1' be so that $B(e_1, e_1') = 1$. Let

$$V = \{z \in \mathbb{R}^n \mid B(z,e_1) = B(z,e_1') = 0\} \quad .$$

Then clearly,

$$\mathbb{R}^n = \mathbb{R}e_1 \oplus \mathbb{R}e_1' \oplus V$$

and $B_{V \times V}$ is non-degenerate. Also

$$B(\mathbb{R}e_1 + \mathbb{R}e_1', V) = 0 \quad .$$

Now use the inductive hypothesis on V. Q.E.D.

Later we will need one more fact about the relationship between $i(X)$, $\mathcal{L}_X$ and d.

(VI) $\mathcal{L}_X(i(Y)\eta) - i(Y)\mathcal{L}_X\eta = i_{[X,Y]}\eta$

for $\eta \in \mathcal{D}^k$, $X,Y \in \mathcal{X}(M)$.

3. THE CANONICAL VOLUME FORM OF A SYMPLECTIC MANIFOLD

Let (M,ω) be a symplectic manifold.

Lemma 3.1. If $\{x_1,\ldots,x_k, y_1,\ldots,y_k\}$ is a system of canonical coordinates on U an open subset of M, then

$$\underbrace{\omega \wedge \cdots \wedge \omega}_{k} = k!(-1)^{[k/2]} dx_1 \wedge \cdots \wedge dx_k \wedge dy_1 \wedge \cdots \wedge dy_k$$

($[k/2] = \ell$ if $k = 2\ell$ or $2\ell + 1$.)

Proof.

$$\underbrace{\omega \wedge \cdots \wedge \omega}_{k} = \sum_{i_1,\ldots,i_k} dx_i \wedge dy_i \wedge \cdots \wedge dx_{i_k} \wedge dy_{i_k}$$

$$= \sum_{i_1,\ldots,i_k \text{ distinct}} dx_{i_1} \wedge dy_{i_1} \wedge \cdots \wedge dx_{i_k} \wedge du_{i_k}$$

$$= k!\ dx_1 \wedge dy_1 \wedge dx_2 \wedge dy_2 \wedge \cdots \wedge dx_k \wedge dy_k$$

Suppose that we have shown that

$$[-1]^{\left[\frac{k-1}{2}\right]} dx_1 \wedge dy_1 \wedge \cdots \wedge dx_{k-1} \wedge dy_{k-1}$$

$$= dx_1 \wedge \cdots \wedge dx_{k-1} \wedge dy_1 \wedge \cdots \wedge dy_{k-1} \quad .$$

Then

$$dx_1 \wedge dy_1 \wedge dx_2 \wedge dy_2 \wedge \cdots \wedge dx_k \wedge dy_k$$

$$= (-1)^{\left[\frac{k-1}{2}\right]} dx_1 \wedge dy_1 \wedge dx_2 \wedge \cdots \wedge dy_2 \wedge \cdots \wedge dy_k$$

$$= (-1)^{\left[\frac{k-1}{2}\right] + k - 1} dx_1 \wedge \cdots \wedge dx_k \wedge dy_1 \wedge \cdots \wedge dy_k$$

Clearly

$$\left[\frac{k-1}{2}\right] + k-1 \equiv \left[\frac{k}{2}\right] \mod 2 \quad .$$

Q.E.D.

Set

$$\Omega = \frac{(-1)^{[k/2]}}{k!} \underbrace{\omega \wedge \cdots \wedge \omega}_{k}$$

Then Ω is called the <u>Pfaffian of</u> ω or <u>canonical volume</u> form of (M,ω).

4. THE AUTOMORPHISM GROUP OF A SYMPLECTIC MANIFOLD

Let (M,ω) be a symplectic manifold. We denote by $\mathrm{Aut}(M,\omega)$ the set of all diffeomorphisms $\phi\colon M \to M$ so that $\phi^*\omega = \omega$. We set $\mathrm{aut}(M,\omega)$ equal to the space of all vector fields X on M so that $L_X\omega = 0$.

If $X \in \mathrm{aut}(M,\omega)$ and if ϕ_t is the local one-parameter group of diffeomorphisms generated by X then

$$\frac{d}{dt}\,\phi_t^*\omega = \phi_t^*(L_X\omega) = 0 \quad .$$

Hence

$$\phi_t^*\omega = \phi_0^*\omega = \omega \quad .$$

We may thus look upon $\mathrm{aut}(M,\omega)$ as the "Lie algebra of $\mathrm{Aut}(M,\omega)$".

Since

$$[L_X, L_Y] = L_{[X,Y]}$$

for $X,Y \in \mathscr{X}(M)$

it is clear that $aut(M,\omega)$ is closed under the bracket operation and hence is a Lie algebra.

Definition 4.1. If $f \in C^\infty(M)$, define $X_f \in \mathscr{X}(M)$ as usual by

$$\omega(X_f, v) = v \cdot f$$

for $v \in T(M)_p$.

Then X_f is called the Hamiltonian vector field associated with f. A vector field X such that there exists $f \in C^\infty(M)$ so that $X = X_f$ is called a Hamiltonian vector field.

Lemma 4.2. If $f \in C^\infty(M)$ then $X_f \in aut(M,\omega)$.

Proof. If $v,\omega \in T(M)_p$ let $h,g \in C^\infty(M)$ be such that

$$(X_h)_p = v , \quad (X_g)_p = \omega .$$

We must show

$$(L_{X_f}\omega)(v,\omega) = 0 .$$

But

$$\begin{aligned}(L_{X_f}\omega)(X_g,X_h) &= X_f\omega(X_g,X_h) - \omega([X_f,X_g],X_h) - \omega(X_g,[X_f,X_h]) \\ &= 0\end{aligned}$$

by Lemma 1.5.

Lemma 4.3. If $\theta \in \mathcal{D}^1(M;\mathbb{R})$, define $X_\theta \in \mathcal{X}(M)$ by

$$i(X_\theta)\omega = \theta .$$

Then $X_\theta \in \mathrm{aut}(M,\omega)$ if and only if $d\theta = 0$. The map

$$\mathrm{Ker}\ d \mapsto \mathrm{aut}(M,\omega)$$

$$\theta \to X_\theta$$

is a bijection.

Proof. If $X_\theta \in \mathrm{aut}(M,\omega)$, then

$$\begin{aligned} 0 = L_{X_\theta}\omega &= d(i(X_\theta)\omega) + i(X_\theta)d\omega \\ &= d(i(X_\theta)\omega) \\ &= d\theta \end{aligned}$$

If $d\theta = 0$, then the same string of equalities proves that $X_\theta \in \mathrm{aut}(M,\omega)$.

If $X_\theta = 0$, then clearly $\theta = 0$. Suppose that $X \in \mathrm{aut}(M,\omega)$. Then, set

$$\theta_X = i(X)\omega .$$

Then, clearly,

$$X_{\theta_X} = X .$$

Q.E.D.

Notice that this notation gives

$$X_{df} = X_f \quad .$$

Let

$$\mathrm{Ham}(M,\omega) = \{X_f \mid f \in C^\infty(M)\}$$

<u>Lemma 4.4</u>. $\mathrm{Ham}(M,\omega) \supset [\mathrm{aut}(M,\omega), \mathrm{aut}(M,\omega)]$.

<u>Proof</u>. Let

$$X,Y \in \mathrm{aut}(M,\omega) \quad , \qquad Z = [X,Y] \quad .$$

Then

$$L_X i(Y)\omega = d(i(X)i(Y)\omega) \quad .$$

Now

$$d(i(X)i(Y)\omega)(U) = U\cdot\omega(Y,X) \quad .$$

Now, (VI), Section 2 says that

$$i([X,Y])\omega = L_X(u(Y)\omega) \quad .$$

Thus, we see

$$\theta_{[X,Y]} = d(\omega(Y,X)) \quad . \qquad \text{Q.E.D.}$$

<u>Note</u>. The de Rham theorem and Lemma 4.3 imply that if $H^1(M;\mathbb{R}) = 0$ then $\mathrm{Ham}(M,\omega) = \mathrm{aut}(M,\omega)$.

5. SYMPLECTIC MANIFOLDS WITH INTEGRAL SYMPLECTIC FORMS (PRE-QUANTIZATION)

We assume that (M,ω) is a symplectic manifold and that

$$[\omega] \xrightarrow{\text{d.R.}} i(H^2(M,\mathbb{Z})) \quad .$$

Let $L \xrightarrow{p} M$ be a line bundle over M, ∇ a connection on L so that $2\pi i\omega = \text{curv}(L,\nabla)$. We have seen that (L,∇) exists. We also assume that there exists a hermitian structure h on L which is ∇ invariant. This is no assumption if $H^1(M,\mathbb{R}) = 0$ (see Lemma I.3.10.)

Let $\theta \in \mathcal{D}^1(L^\times,\mathbb{C})$ be the connection form of (L,∇). Let $\underline{e}(L,\nabla)$ be the space of all $X \in \mathscr{X}(L^\times)$ satisfying

1) $L_X\theta = 0$

2) $X\cdot|h|^2 = 0 \qquad (|h|^2(v) = h(v,v))$.

3) If $c \in \mathbb{C}^\times$, then $c_{*s}X_s = X_{s\cdot c}$.

Note. Once we know h exists, it is unique up to scalar multiple. Thus, the notation $\underline{e}(L,\nabla)$ is appropriate.

Lemma 5.1. Let $X \in \mathscr{X}(M)$. Define $\tilde{X}_s \in T(L^\times)_s$ to be the unique element of H_s (the horizontal space,

$$H_s = \{v \in T(L^\times)_s \,|\, \theta_s(v) = 0\} \;),$$

such that

$$p_{*s}(\tilde{X}_s) = X_s \quad .$$

Then

a) $\tilde{X} \in \mathscr{X}(L^\times)$

b) $\tilde{X}$ generates a global one-parameter group of diffeomorphisms of $L^\times$ if and only if X generates a global one-parameter group of diffeomorphisms of M.

<u>Proof</u>. Let ϕ_t be the local one-parameter group of diffeomorphisms generated by X. Let

$$s_0 \in L^\times \ , \qquad p(s_0) = x \ .$$

Let

$$t \mapsto \tilde{\phi}_t(s_0)$$

be the (unique) parallel cross-section over $t \mapsto \phi_t(x)$ satisfying $\tilde{\phi}_0(s_0) = s_0$. Then $\tilde{\phi}_t$ defines a local one-parameter group of diffeomorphisms of L^X. Clearly $\tilde{\phi}$ is generated by $\tilde{X}$.

If ϕ is a global one-parameter group of diffeomorphisms of M, then clearly $\tilde{\phi}_t$ is defined for all t. Hence $\tilde{\phi}$ is a global one-parameter group of diffeomorphisms.

If $\tilde{\phi}_t$ is defined for all t then since

$$p(\tilde{\phi}_t(s_0)) = \phi_t(x) \ , \qquad (p(s_0) = x),$$

ϕ_t is defined for all t. Hence X generates a global one-parameter group of diffeomorphisms. Q.E.D.

If $\Phi \in C^{\infty}(M;\mathbb{C})$, define

$$\eta(\Phi) \in \mathscr{X}(L^{\times})$$

to be the vector field generating the one-parameter group of diffeomorphisms of $L^{\times}$ given by

$$\psi_t(s) = se^{-2\pi it\Phi\circ p}$$

clearly

$$\eta(\Phi)_s = -\beta(s)_{*1}(2\pi i\Phi(p(s))) \quad .$$

If

$$\Phi \in C^{\infty}(M;\mathbb{C}) \ , \qquad X \in \mathcal{X}(M)$$

define

$$\eta(\Phi,X) = \eta(\Phi) + \tilde{X} \quad .$$

Clearly, $\eta(\Phi,X) \in \mathscr{X}(L^{\times})$.

Lemma 5.2. a) $\eta(\Phi,X)$ generates a global one-parameter group of diffeomorphisms of $L^{\times}$ if and only if X generates a global one-parameter group of diffeomorphisms of M.

b) $\mathcal{L}_{\eta(\Phi,X)}\theta = 2\pi ip^*(\beta_X - d\Phi)$, $\quad \beta_X = i(X)\omega$.

Proof. a) Suppose that X generates a global one-parameter group ϕ_t. Let $\tilde{\phi}_t$ be as in Lemma 5.1. Define

$$g(t,p) = e^{-2\pi i \int_0^t \Phi(\phi_t(p))\,dt} .$$

Then

$$t \to g(t,p(s))\tilde{\phi}_t(s)$$

defines a global one-parameter group for $\eta(\Phi,X)$. Since $p_*\eta(\Phi,X) = X$, a) follows.

b) $$L_{\eta(\Phi,X)}\theta = i(\eta(\Phi,X))d\theta + d(i(\eta(\Phi,X))\theta) .$$

Now,

$$\begin{aligned} i(\eta(\Phi,X))\theta &= \theta(X^\vee + \eta(\Phi)) \\ &= \theta(\eta(\Phi)) = -2\pi i\Phi . \end{aligned}$$

Hence

$$d(i(\eta(\Phi,X))\theta) = -2\pi i p^* d\Phi .$$

Now,

$$d\theta = 2\pi i p^*(\omega) .$$

Hence,

$$\begin{aligned} i(\eta(\Phi,X))d\theta &= i(\eta(\Phi) + \tilde{X})d\theta \\ &= 2\pi i(i(\eta(\Phi) + \tilde{X}))p^*\omega \\ &= (2\pi i)(i(\eta(\Phi))p^*\omega) + 2\pi i p^*(i(X)\omega) . \end{aligned}$$

Clearly, $p_*\eta(\Phi) = 0$. We therefore have the desired formula.

Lemma 5.3. $\eta(\Phi,X) \in \underline{e}(L,\nabla)$ if and only if

1) Φ is real valued and

2) $\beta_X = d\Phi$.

Proof. Suppose that $\eta(\Phi,X)|h|^2 = 0$. Then

$$\eta(\Phi)|h|^2 + \tilde{X}|h|^2 = 0 .$$

But h is ∇-invariant. Hence

$$\frac{dh}{dt}(\tilde{\phi}_t(s), \tilde{\phi}_t(s)) = 0 .$$

Hence $\tilde{X}|h|^2 = 0$.

Now,

$$\eta(\Phi_s)|h|^2 = \frac{d}{dt}_{t=0} h\left(e^{-2\pi it\Phi(p(s))} s, e^{-2\pi it\Phi(p(s))} s\right)$$

$$= 2\pi i(\overline{\Phi} - \Phi)(p(s))|h|^2(s) .$$

This implies that

$$\eta(\Phi_s)|h|^2 = 0$$

if and only if $\Phi = \overline{\Phi}$ or $\Phi \in C^\infty(M;\mathbb{R})$.

Finally,

$$L_{\eta(\Phi,X)}\theta = 0 \Rightarrow p^*(\beta_X - d\Phi) = 0$$

if and only if

$$\beta_X = d\Phi . \qquad \text{Q.E.D.}$$

<u>Lemma 5.4</u>. If $X \varepsilon \mathscr{X}(L^{\times})$ and $c_{*s}X_s = X_{s\cdot c}$ for all $c \varepsilon \mathbb{C}$ then

$$X = \eta(\Phi, Z)$$

with $\Phi \varepsilon C^{\infty}(M;\mathbb{C})$ and $Z \varepsilon \mathscr{X}(M)$.

Set

$$Z = \overset{\vee}{X} \quad .$$

Then $[Z,W]^{\vee} = [Z^{\vee},W^{\vee}]$.

<u>Proof</u>. Let X_s^H be the horizontal projection of X_s for $s \varepsilon L^{\times}$. Set

$$Z_X = P_{*s}X_s^H \quad .$$

We must show that Z_X is well defined and that $X \mapsto Z_X$ is a vector field on M.

To see that Z_X is well defined we must only note that

$$c_{*s}\cdot H_s = H_{s\cdot c} \quad .$$

Thus

$$P_{*s.c}\, X_{s\cdot c}^H = P_{*s}X_s^H$$

for $c \varepsilon \mathbb{C}^{\times}$.

Since $P^{-1}(X) = X\cdot\mathbb{C}^{\times}$, Z_X is well defined. Let ϕ_t be the local one-parameter group of diffeomorphisms generated by X. Then

$$\phi_t(s\cdot c) = \phi_t(s)\cdot c$$
$$\text{for} \quad c \in \mathbb{C}^\times$$

by the uniqueness theorem. Hence, if we set

$$\check{\phi}_t(X) = p(\phi_t(s))$$
$$\text{for} \quad s \in p^{-1}(X)$$

$\check{\phi}_t$ defines a local one-parameter group of diffeomorphisms of M. Clearly,

$$Z_X f = \frac{d}{dt}\bigg|_{t=0} f(\check{\phi}_t(X)) \quad .$$

Hence, $Z \in \mathscr{X}(M)$.

Now $X = X^H + X^\vee$. Since $c_{*s}X_s = X_{s\cdot c}$, $c_{*s}X_s^H = X_{s\cdot c}^H$ we see $c_{*s}X_s^\vee = X_{s\cdot c}^\vee$. But

$$X_s^\vee = \beta(s)_{*1}(u(s)) \ ,$$
$$u \in C^\infty(L^\times, \mathbb{C}) \quad .$$

Indeed, $u(s) = \theta(X_s)$. Now $c^*\theta = \theta$. Hence, $u(s\cdot c) = u(s)$. This implies that

$$u(s) = -2\pi i \Phi(p(s))$$
$$\text{for} \quad \Phi \in C^\infty(M;\mathbb{C}) \quad .$$

Hence we see that $X = \eta(\Phi, Z)$.

To prove the last assertion we note that if X,Y satisfy $c_*X = X$, $c_*Y = Y$ for $c \in \mathbb{C}^\times$, then if ϕ_t is the local one-parameter group of diffeomorphisms generated by X

$$P_{*s}\left(\phi_{-t*\phi_t(s)} Y_{\phi_t(s)}\right) = \check{\phi}_{-t*\check{\phi}_t(p(s))} P_{*\phi_t(s)} Y_{\phi_t(s)}$$

This is an easy consequence of invariance. Taking the derivative at $t = 0$ gives the last assertion of the lemma.

Lemma 5.5. $[\eta(\Phi_1,X_1),\eta(\Phi_2,X_2)] = \eta(\omega(X_1,X_2) + X_1\Phi_2 - X_2\Phi_1,[X_1,X_2])$

Proof. We first note that

$$\begin{aligned} c_*[\eta(\Phi_1,X_1),\eta(\Phi_2,X_2)] &= [c_*\eta(\Phi_1,X_1),c_*\eta(\Phi_2,X_2)] \\ &= [\eta(\Phi_1,X_1),\eta(\Phi_2,X_2)] \ . \end{aligned}$$

Hence Lemma 4.4 implies

$$[\eta(\Phi_1,X_1],\eta(\Phi_2,X_2)] = \eta(\Phi,X)$$

with $\Phi \in C^\infty(M;\mathbb{C})$, $X \in \mathscr{X}(M)$.

Clearly $\eta(\psi,Z)^\vee = Z$. Hence,

$$X = [X_1,X_2]$$

by Lemma 5.4.

Now if

$$\eta_i = \eta(\Phi_i, X_i) \quad ,$$

$$i = 1,2 \quad ,$$

then

$$\theta([\eta_1,\eta_2]) = \theta(\eta(\Phi,X)) = -2\pi i\Phi \quad .$$

Hence,

$$-\frac{1}{2\pi i}\,\theta([\eta_1,\eta_2]) = \Phi$$

But

$$d\theta(\eta_1,\eta_2) = \eta_1\cdot\theta(\eta_2) - \eta_2\cdot\theta(\eta_1) - \theta([\eta_1,\eta_2]) \quad .$$

$$d\theta = 2\pi i p^*\omega \quad .$$

Hence,

$$\begin{aligned}
-\frac{1}{2\pi i}\,\theta([\eta_1,\eta_2]) &= (p^*\omega)(\eta_1,\eta_2) - \frac{1}{2\pi i}\,\eta_1\theta(\eta_2) \\
&\quad + \frac{1}{2\pi i}\,\eta_2\theta(\eta_1) \\
&= \omega(X_1,X_2) + \eta_1 p^*\Phi_2 - \eta_2 p^*\Phi_1 \\
&= \omega(X_1,X_2) + p_*\eta_1\cdot\Phi_2 - p_*\eta_2\cdot\Phi_1 \\
&= \omega(X_1,X_2) + X_1\Phi_2 - X_2\Phi_1 \quad .
\end{aligned}$$

Q.E.D.

Corollary 5.6. $[\eta(\Phi_1, X_{\Phi_1}), \eta(\Phi_2, X_{\Phi_2})] = \eta(X_{\Phi_1}\Phi_2, [X_{\Phi_1}, X_{\Phi_2}])$

Proof. $\omega(X_{\Phi_1}, X_{\Phi_2}) = X_{\Phi_2}\Phi_1$.

Definition 5.7. If $\Phi \in C^\infty(M;\mathbb{R})$, set $\tilde{\delta}(\Phi) = -\eta(\Phi, X_\Phi)$.

Theorem 5.8 (Kostant [16]). a) The following sequence is an exact sequence of Lie algebra homomorphisms:

$$0 \to \mathbb{R} \xrightarrow{\eta} \underline{e}(L,\nabla) \xrightarrow{(X \mapsto \check{X})} \mathrm{Hom}(M,\omega) \to 0$$

($\mathbb{R}$ is identified with the constant functions on M.)

b) The following is a commutative diagram of Lie algebra homomorphisms with top and bottom exact.

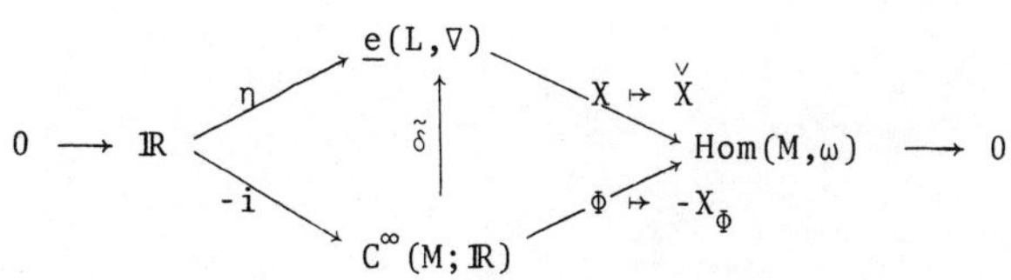

Proof. a) We note

$$X_{\{\Phi_1,\Phi_2\}} = -[X_{\Phi_1}, X_{\Phi_2}] \ .$$

If $Z \in \underline{e}(L,\nabla)$, then Lemma 5.4 implies $Z = \eta(\Phi, X)$. Lemma 5.3 implies $\Phi \in C^\infty(M;\mathbb{R})$ and $i(X)\omega = d\Phi$. Hence, $X = X_\Phi$. Hence, $Z = \eta(\Phi, X_\Phi)$. $X_\Phi = 0$ if and only if $d\Phi = 0$, that is $\Phi \in \mathbb{R}$. This proves a) in light of Lemma 5.4.

To prove b) we note that

$$[\tilde{\delta}(\Phi_1),\ \tilde{\delta}(\Phi_2)] = \eta(X_{\Phi_1}\Phi_2, [X_{\Phi_1}, X_{\Phi_2}])$$

$$= -\eta(\{\Phi_1,\Phi_2\},\ X_{\{\Phi_1,\Phi_2\}})$$

$$= \tilde{\delta}(\{\Phi_1,\Phi_2\}) \quad .$$

Thus, $\tilde{\delta}$ is a Lie algebra homomorphism. Clearly $\Phi \mapsto -X_\Phi$ is a Lie algebra homomorphism of $C^\infty(M;\mathbb{R})$ onto $\mathrm{Hom}(M,\omega)$ with kernel $\mathbb{R}$.

Clearly we have

$$\tilde{\delta}(i(x)) = -\eta(x,0) = -\eta(c) \quad .$$

$$\tilde{\delta}(\Phi)^\vee = -X_\Phi$$

proving the commutivity .

Definition 5.9. If $s \in C^\infty(M;L)$, define $\tilde{s} \in C^\infty(L^\times,\mathbb{C})$ by $\tilde{s}(x)x = s(p(x))$.

Lemma 5.10. If $X \in \mathscr{X}(M)$, then

$$\tilde{X}\tilde{s} = \tilde{u}$$

with $u = \nabla_X s$.

<u>Proof</u>. Let $\tilde{\phi}_t$ be the local one-parameter group of diffeomorphisms generated by $\tilde{x}$. Then if $s \in C^\infty(M;L)$,

$$s(\phi_t x) = h(\phi_t x)\tilde{\phi}_t(s(x)) \quad .$$

Also,

$$\begin{aligned}\tilde{s}(\tilde{\phi}_t(s(x)))\tilde{\phi}_t(s(x)) &= s(p(\tilde{\phi}_t(s(x)) \\ &= s(\phi_t(x)) \quad .\end{aligned}$$

Hence,

$$h(\phi_t x) = \tilde{s}(\tilde{\phi}_t(s(x)) \quad .$$

Now,

$$\frac{D}{dt} s(\phi_t x) = \frac{d}{dt} h(\phi_t x)\tilde{\phi}_t(s(x))$$

Since

$$\frac{D}{dt} \tilde{\phi}_t(s(x)) = 0 \quad .$$

Hence,

$$\begin{aligned}\nabla_X s &= \left(\frac{d}{dt}_{t=0} \tilde{s}(\tilde{\phi}_t(s(x))\right)\cdot s(x) \\ &= (\tilde{x}\tilde{s})(s(x))s(x) \quad .\end{aligned}$$

This proves the lemma.

Lemma 5.11. If $\Phi \in C^\infty(M;\mathbb{C})$, $X \in \mathscr{X}(M)$ and $s \in C^\infty(M;L)$, then

$$\eta(\Phi,X)\tilde{s} = \tilde{u}$$

$$\text{with } u = (\nabla_X + 2\pi i\Phi)s \quad .$$

Proof. $\eta(\Phi,X) = \eta(\Phi) + \tilde{X}$. In light of Lemma 5.10 we need only show that $\eta(\Phi)\tilde{s} = \tilde{u}$ with $u = 2\pi i\Phi s$.

But

$$(\eta(\Phi)\tilde{s})(z) = \frac{d}{dt}\Big|_{t=0} \tilde{s}(e^{-2\pi it\Phi(p(z))}z)$$

$$= \frac{d}{dt}\Big|_{t=0} e^{2\pi it\Phi(p(z))}\tilde{s}(z)$$

(since $\tilde{s}(cz)cz = s(p(z))$. Hence,

$$\tilde{s}(cz) = c^{-1}\tilde{s}(z) \qquad \text{for } z \in L^\times, \ c \in \mathbb{C}^\times$$

$$= 2\pi i\Phi(p(z))\tilde{s}(z)$$

$$= (2\pi i\Phi s)^{\sim}(z) \quad .$$

Definition 5.12. Define for $\Phi \in C^\infty(M;\mathbb{R})$ and $s \in C^\infty(M;\mathbb{R})$,

$$\delta(\Phi)s = -(\nabla_{X_\Phi} + 2\pi i\Phi)s \quad .$$

Lemma 5.13.

a) $\delta(\{\Phi,\psi\}) = \delta(\Phi)\delta(\psi) - \delta(\psi)\delta(\Phi)$

for $\Phi,\psi \in C^\infty(M;\mathbb{R})$.

b) $(\delta(\Phi)s)^\sim = \tilde{\delta}(\Phi)\tilde{s}$

c) Define for f or $g \in C_c^\infty(M;L)$

$$(f,g) = \int_M h(f(x),g(x))\Omega$$

(Ω the Pfaffian of ω.) Then,

$$(\delta(\Phi)f,g) = -(f,\delta(\Phi)g)$$

for f or $g \in C_c^\infty(M;L)$, $\Phi \in C^\infty(M;\mathbb{R})$.

<u>Proof</u>. b) is just a restatement of Lemma 5.11. a) follows from the fact that $\tilde{\delta}$ is a Lie algebra homomorphism and b).

To prove c) we note

$$\begin{aligned} h(\delta(\Phi)f,g) &= -h((\nabla_{X_\Phi} + 2\pi i\Phi)f,g) \\ &= -h(\nabla_{X_\Phi} f,g) - 2\pi i\Phi h(f,g) \\ &= -X_\Phi h(f,g) + h(f,\nabla_{X_\Phi} g) - 2\pi i\Phi h(f,g) \\ &= -X_\Phi h(f,g) + h(f,(\nabla_{X_\Phi} + 2\pi i\Phi)g) \quad . \end{aligned}$$

Thus we need only note that if $u \in C_c^\infty(M;\mathbb{C})$,

$$\int_M (X_\Phi u)\Omega = -\int_M u(L_{X_\Phi}\Omega) \quad .$$

But

$$L_{X_\Phi}\omega = 0 \quad .$$

Hence, the result follows.

<u>Definition 5.14</u>. The map

$$\delta\colon C^\infty(M;\mathbb{R}) \to \mathrm{End}\ (C^\infty(M;L))$$

is called <u>pre-quantization</u>.

<u>Note</u>. If $L^2(M;L)$ is the completion of $C_c^\infty(M;L)$ relative to $(\ ,\ )$, then $\delta(\Phi)$ is a densely defined formally skew-adjoint operator on $L^2(M;L)$. Prequantization "turns real valued functions into skew-Hermitian operators on a Hilbert space."

6. EXAMPLE. $T^*(\mathbb{R}^n)$.

Let $M = T^*(\mathbb{R}^n)$ with the canonical symplectic structure ω; see Example 1.2. Since

$$0 - H^2(M,\mathbb{Z}) = H^2(M,\mathbb{R}) = H^2(M,\mathbb{C}) \quad ,$$

ω is integral. We take $y_1,\ldots,y_n$ to be the $\mathbb{R}^n$ coordinates

$x_1,\dots,x_n$ the $T^*(\mathbb{R}^n)_p$ coordinates (i.e., if $x \in T^*(\mathbb{R}^n)_p$ then $\lambda = \Sigma\, x_i\, dy_{i_p}$). Then

$$\omega = \sum_{i=1}^{n} ds_i \wedge dy_i \quad .$$

Using these coordinates we may look upon $T^*(\mathbb{R}^n)$ as $\mathbb{R}^{2n}$.

$$\Omega = dx_1 \wedge \cdots \wedge dx_n \wedge dy_1 \wedge \cdots \wedge dy_n$$

(see Section 4) the usual volume form on $\mathbb{R}^{2n}$.

Take L to be the trivial line bundle $M \times \mathbb{C}$ (there are no others). A cross-section of L will be identified with a C^∞ function from $\mathbb{R}^{2n}$ to $\mathbb{C}$. Let

$$s(x) \equiv 1 \quad ,$$
$$x \in \mathbb{R}^{2n} \quad .$$

Then $s \in C^\infty(M;L)$. Set

$$\theta = \sum x_i\, dy_i \quad .$$

Then, $\omega = d\theta$.

Define

$$\nabla_v s = 2\pi i \theta(v) s(x)$$
$$\text{for } v \in T(M)_x \quad .$$

Then

$$\nabla_v(fs) = (v\cdot f)s + f(x)\nabla_v s$$

defines a connection on L. Now this just says

$$\nabla_v f = v\cdot f + \theta(v)f(x) \quad ,$$

$$x \in M \ , \quad v \in T(M)_x \quad .$$

Hence, if $f \in C^\infty(M;\mathbb{C}) = C^\infty(M;L)$, then

a) $$\nabla_{\frac{\partial}{\partial x_i}} \cdot f = \frac{\partial f}{\partial x_i}$$

b) $$\nabla_{\frac{\partial}{\partial y_i}} \cdot f = \frac{\partial f}{\partial y_i} + 2\pi i x_i f$$

It is easily checked then that $h_{(x,y)}(z,w) = z\bar{w}$ defines a ∇-invariant Hermitian structure on L. If $\Phi \in C^\infty(\mathbb{R}^{2n};\mathbb{R})$ then

c) $$X_\Phi = -\sum \frac{\partial\Phi}{\partial x_i}\frac{\partial}{\partial y_i} + \sum \frac{\partial\Phi}{\partial y_i}\frac{\partial}{\partial x_i}$$

since $x_1,\ldots,x_n,\ y_1,\ldots,y_n$ is a system of canonical coordinates.

Now

$$\begin{aligned}\delta(\Phi) &= -(\nabla_{X_\Phi} + 2\pi i\Phi)\\ &= \sum \frac{\partial\Phi}{\partial x_i}\left(\frac{\partial}{\partial y_i} + 2\pi i x_i\right) - \sum \frac{\partial\Phi}{\partial y_i}\frac{\partial}{\partial x_i} - 2\pi i\Phi\end{aligned}$$

$$= - X_\Phi + 2\pi i \left(\sum x_i \frac{\partial \Phi}{\partial x_i} - \Phi \right)$$

We now study an interesting class of functions. Let, for $(v,w,t) \in \mathbb{R}^n \times \mathbb{R}^n \times \mathbb{R}$

$$\psi(v,w,t)(x,y) = \sum v_i x_i + \sum w_i y_i + t \quad .$$

Then

$$\psi\colon \mathbb{R}^n \times \mathbb{R}^n \times \mathbb{R} \to C^\infty(\mathbb{R}^{2n})$$

1) $$\{\psi(v,w,t), \psi(v',w',t')\} \equiv \langle v,w'\rangle - \langle w,v'\rangle$$

(here $\langle v,w\rangle = \Sigma\, v_i w_i$) .

Let h_n denote $\mathbb{R}^n \times \mathbb{R}^n \times \mathbb{R}$ with

$$[(v,w,t), (v',w',t')] = (0,0,\langle v,w'\rangle - \langle w,v'\rangle) \quad .$$

Then, by (1) h_n is a Lie algebra and $\psi\colon h_n \to C^\infty(\mathbb{R}^{2n})$ is a Lie algebra monomorphism.

A direct computation gives

2) $$\delta(\psi(v,w,t)) = \sum v_i \frac{\partial}{\partial y_i} - \sum w_i \frac{\partial}{\partial x_i} - 2\pi i \left(\sum y_i w_i + t \right)$$

We will now give a "preview" of quantization. We wish to represent h_n by operators on $\mathbb{R}^n$ (rather than $\mathbb{R}^{2n}$). Noting that we can imbed $C^\infty(\mathbb{R}^n)$ into $C^\infty(\mathbb{R}^{2n})$ by identifying

$f \in C^\infty(\mathbb{R}^n)$ with $\tilde{f} \in C^\infty(\mathbb{R}^{2n})$

$$\frac{\partial \tilde{f}}{\partial x_i} = 0 , \qquad i = 1,\ldots,n \quad .$$

That is, if $p\colon \mathbb{R}^{2n} \to \mathbb{R}^n$, $p(x,y) = y$, then $\tilde{f} = f \circ p$.

Now,

$$(\delta \circ \psi)(v,w,t)\cdot \tilde{f} = \sum v_i \frac{\partial}{\partial y_i} \tilde{f} - 2\pi i(<v,w> + t)\tilde{f} \quad .$$

Hence,

$$(\delta \circ \psi)(v,w,t) C^\infty(\mathbb{R}^n) \circ p \subset C^\infty(\mathbb{R}^n) \circ p \quad .$$

This induces an operator on $C^\infty(\mathbb{R}^n)$.

$$q(v,w,t)f = \sum v_i \frac{\partial}{\partial y_i} f - 2\pi i(<v,w> + t)f \quad .$$

One checks easily that

$$\int_{\mathbb{R}^n} (q(v,w,t)f)(y)\, \overline{g(y)}\, dy = - \int_{\mathbb{R}^n} f(y)\, (\overline{q(v,w,t)g)(y)}\, dy$$

for, say, $f,g \in C_c^\infty(\mathbb{R}^n)$.

q gives "quantization" in this simple case. We note that if we had used $X_{\psi(v,w,t)}$ rather than $\delta(\psi(v,w,t))$, then we could not "cut down" to $\mathbb{R}^n$.

Chapter 3

HOMOGENEOUS SYMPLECTIC MANIFOLDS

In this chapter we use the theory developed in Chapter 2 to study symplectic actions of Lie groups on symplectic manifolds. Our development follows Kostant [16] with minor changes. Included in this chapter is a "recipe" for all simply connected, homogeneous, symplectic manifolds. It is of interest that the recipe for $T^*(\mathbb{R}^n)$ with the canonical symplectic structure is given in terms of the natural action of the Heisenberg group. Also in this chapter we give Kostant's criterion for the integrality of the symplectic structure on a homogeneous symplectic manifold. Most of the above mentioned material can also be found in Souriau [24].

We also include an appendix to show how the results of this chapter can be modified to apply to Tori. As a bonus, we find that the classical theta functions (see Weil [30]) are "built into the theory."

1. SYMPLECTIC G-SPACES

Let G be a Lie group and let (M,ω) be a symplectic manifold. If G acts on M by diffeomorphisms (the action of G on M is denoted $g,x \mapsto g\cdot x$, $g \in G$, $x \in M$) so that

$$g^*\omega = \omega$$

for $g \in G$

then (M,ω) is called a symplectic G-space. (M,ω) is called a homogeneous symplectic G-space if G (in addition) acts transitively.

In what follows, we will assume that G is connected.

If M is a manifold and if G is a Lie group acting on M, and if $\mathfrak{g}$ is the Lie algebra of G, $\exp: \mathfrak{g} \to G$ the exponential map of G (cf. Warner [27]), we define

$$\sigma: \mathfrak{g} \to \mathscr{X}(M)$$

by

$$(\sigma(X)f)(p) = \frac{d}{dt} f(\exp tX\cdot p)|_{t=0} .$$

We will also use the notation $\sigma(X) = X^*$. We note that

$$\sigma([X,Y]) = -[\sigma(X), \sigma(Y)] .$$

Definition 1.1. A symplectic G-space (M,ω) is called strongly symplectic if $\sigma(\mathfrak{g}) \subset \mathrm{Ham}(M,\omega)$ (see 2.4).

Lemma 1.2. If (M,ω) is a G-space and $\mathfrak{g} = [\mathfrak{g},\mathfrak{g}]$ or $H'(M;\mathbb{C}) = 0$, then (M,ω) is strongly symplectic.

Proof. This is an immediate consequence of Lemmas 2.4.3 and 2.4.4.

If (M,ω) is symplectic, we have the exact sequence of Lie algebras

$$0 \longrightarrow \mathbb{R} \xrightarrow{\ i\ } C^{\infty}(M,\mathbb{R}) \xrightarrow{\ \phi \to -X_{\phi}\ } \mathrm{Ham}(M,\omega) \longrightarrow 0$$

Suppose (M,ω) is a strongly symplectic G-space. Then a <u>lift</u> of $\phi\colon \mathfrak{g} \to \mathrm{Ham}(M,\omega)$ is a Lie algebra homomorphism

$$\lambda\colon \mathfrak{g} \to C^{\infty}(M;\mathbb{R})$$

so that

$$\begin{array}{ccccccccc} 0 & \longrightarrow & \mathbb{R} & \longrightarrow & C^{\infty}(M;\mathbb{R}) & \longrightarrow & \mathrm{Ham}(M,\omega) & \longrightarrow & 0 \\ & & & & & {}_{\lambda}\nwarrow & \uparrow{\scriptstyle -\sigma} & & \\ & & & & & & \mathfrak{g} & & \end{array}$$

commutes.

Let $X_1,\ldots,X_n$ be a basis of $\mathfrak{g}$ and let $\sigma(X_i) = X_{\phi_i}$. Let $\mu_0(X_i) = \phi_i$, and define

$$\mu_0\left(\sum a_i X_i\right) = a_i\mu_0(X_i)$$

$$\text{for } a_i \in \mathbb{R}$$

Then $\mu_0\colon \mathfrak{g} \to C^{\infty}(M;\mathbb{R})$ and $X_{\mu_0(X)} = \sigma(X)$. Thus if μ_0 were a Lie algebra homomorphism then μ_0 would be a lift.

Before we go on we need the notion of Lie algebra cohomology group. Let $\mathfrak{g}$ be a Lie algebra over $\mathbb{R}$. Let $C^k(\mathfrak{g};\mathbb{R})$ be the space of all k-linear, alternating maps $\gamma\colon X^k\mathfrak{g} \to \mathbb{R}$. Define

$$\delta\colon C^k(g;\mathbb{R}) \to C^{k+1}(g;\mathbb{R})$$

by

$$(\delta\gamma)(X_1,\ldots,X_{k+1})$$

$$= \sum_{i<j} (-1)^{i+j}\gamma([X_i,X_j],X_1,\ldots,X_i,\ldots,X_j,\ldots,X_{k+1})$$

The Jacobi identity implies that $\delta^2 = 0$. Set

$$H^k(g;\mathbb{R}) = \frac{\operatorname{Ker}\delta\big|_{C^k(g;\mathbb{R})}}{\delta C^{k-1}(g;\mathbb{R})}$$

$H^k(g;\mathbb{R})$ is called the k-th Lie algebra cohomology group g (with values in the trivial module). See Jacobson [12] for a discussion of Lie algebra cohomology. For our purposes, we need only know that $H^k(g;\mathbb{R})$ is an intrinsic object associated with g.

We return to the situation we were studying. Set, for $X,Y \in g$,

$$\mu(X,Y) = \{\mu_0(X), \mu_0(Y)\} - \mu_0([X,Y]) .$$

Then

$$X_{\mu(X,Y)} = \sigma([X,Y]) - \sigma([X,Y]) = 0 .$$

Hence,

$$\mu(X,Y) \equiv \beta(X,Y) \in \mathbb{R} \quad .$$

Clearly,

$$\beta(X,Y) = -\beta(Y,X) \quad .$$

Lemma 1.3. $\delta\beta = 0$.

Proof. We must only show that

$$-\{\mu_0([X,Y]),\mu_0(Z)\} + \{\mu_0([X,Y]),\mu_0(Y)\} - \{\mu_0([Y,Z]),\mu_0(X)\} = 0$$

for $X,Y,Z \in g$.

But

$$\begin{aligned}
\{\mu_0([X,Y]),\mu_0(Z)\} &= -X_{\mu_0([X,Y])}\mu_0(Z) \\
&= -\sigma([X,Y])\mu_0(Z) \\
&= [\sigma(X),\sigma(Y)]\mu_0(Z) \\
&= [X_{\mu_0(X)},X_{\mu_0(Y)}]\mu_0(Z) \\
&= -X_{\{\mu_0(X),\mu_0(Y)\}}\mu_0(Z) \\
&= \{\{\mu_0(X),\mu_0(Y)\},\mu_0(Z)\}
\end{aligned}$$

Thus the identity follows from Lemma 2.1.5.

Lemma 1.4. If $\beta = d\alpha$ for some $\alpha \in C^1(g,\mathbb{R})$ then there exists a lift λ. In particular, if $H^2(g,\mathbb{R}) = 0$ there is always a lifting.

Proof. If $\beta = d\alpha$ then $\beta(X,Y) = -\alpha([X,Y])$, $\alpha \in g^*$. Define

$$\lambda(X)(p) = \mu_0(X)(p) + \alpha(X) \quad .$$

Then

$$\lambda([X,Y]) - \{\lambda(X),\lambda(Y)\} = \mu_0([X,Y]) + \alpha([X,Y]) - \{\mu_0(X),\mu_0(Y)\}$$

$$= \beta(X,Y) + \alpha([X,Y]) = 0$$

Q.E.D.

If λ_1,λ_2 are two lifts of σ then $\lambda_1(X) - \lambda_2(X)$ is a constant for $X \in g$. Hence $(\lambda_1-\lambda_2)\colon g \to \mathbb{R}$ is a Lie algebra homomorphism. Hence,

$$(\lambda_1 - \lambda_2)|_{[g,g]} \equiv 0 \quad .$$

We have proved

Proposition 1.5. If $H^2(g,\mathbb{R}) = 0$ then there exists a lift of σ. If $H^2(g,\mathbb{R}) = 0$ and $[g,g] = g$ (i.e., $H^1(g,\mathbb{R}) = 0$) then there exists a unique lift of σ.

3. HAMILTONIAN G-SPACES

Definition 2.1. A triple (M,ω,λ) consisting of a strongly symplectic G-space, (M,ω), and a lift λ of σ is called a Hamiltonian G-space.

Example 2.2. Let G be a Lie group with Lie algebra . Let g^* be the dual space of g and let G act on g^* as usual. We have shown that if $f \in g^*$ and $\mathcal{O}_f = G\cdot f$,

$$\omega_x^f(X^*,Y^*) = x([X,Y])$$

then $(\mathcal{O}_f,\omega^f)$ is a symplectic G-space.

Let $\lambda_f\colon g \to C^\infty(\theta_f,\mathbb{R})$ be defined by

$$\lambda_f(X)(p) = p(X) .$$

We compute

$$X_p^*\lambda_f(Y) = \frac{d}{dt}\Big|_{t=0} \lambda_f(Y)(\exp tX\cdot p)$$

$$= \frac{d}{dt}\Big|_{t=0} (\exp tX\cdot p)(Y) = (X\cdot p)(Y)$$

$$= -p([X,Y]) = -\omega_p(X^*,Y^*) = \omega_p(Y^*,X^*) .$$

Thus

$$X_{\lambda_f(Y)} = \sigma(Y) .$$

Also,

$$\{\lambda_f(Y),\lambda_f(X)\}(p) = \omega_p(X_{\lambda_f(Y)}, X_{\lambda_f}(Y))$$

$$= \omega_p(Y^*,X^*) = p([Y,X])$$

$$= \lambda_f([X,Y])(p) .$$

Hence,

$$\{\lambda_f(Y),\lambda_f(X)\} = \lambda_f([Y,X]) .$$

This implies $(\mathcal{O}_f,\omega^f,\lambda_f)$ <u>is a Hamiltonian</u> G-<u>space</u>.

<u>Observation 2.3</u>. We note that if (M,ω) is a strongly simplectic G-space and if

$$\mu_0 : \mathcal{g} \to C^\infty(M,\mathbb{R})$$

satisfies $X_{\mu_0(X)} = \sigma(X)$ but the corresponding β is not of the form $-\delta\alpha$, $\alpha \in \mathcal{g}^*$ we can still extend $\mathcal{g}$ and G to find a lift. Indeed, let

$$\tilde{\mathcal{g}}_\beta = \mathcal{g} \times \mathbb{R}$$

and define

$$[(X,t),(Y,s)] = ([X,Y],\beta(X,Y)) .$$

Then $\tilde{\mathcal{g}}$ is a Lie algebra (the Jacobi identity is the Jacobi identity for $\mathcal{g}$ and $\delta\beta = 0$). Define

$$\lambda((X,t)) = \mu_0(X) + t \quad .$$

Then

$$\begin{aligned}\{\lambda(X,t),\lambda(Y,s)\} &= \{\mu_0(X),\mu_0(Y)\} \\ &= \mu_0([X,Y]) + \beta(X,Y) \\ &= \lambda(([X,Y],\beta(X,Y))) \\ &= \lambda([(X,t),(Y,s)]) \quad .\end{aligned}$$

Let $\tilde{G}_\beta$ be the connected simply connected Lie group with Lie algebra $\tilde{g}$. Let H be the connected subgroup with Lie algebra $\{(0,t) \mid t \in \mathbb{R}\}$. Then $\tilde{G}_\beta/H$ is the simply connected Lie group with Lie algebra g. Let

$$\nu\colon \tilde{G}_\beta/H \to G$$

be the covering map. Let

$$\mu\colon \tilde{G}_\beta \to G$$

be defined by

$$\mu(g) = \nu(gH) \quad .$$

Then $\tilde{G}$ acts on M by $\tilde{g}\cdot p = \mu(\tilde{g})\cdot p$, $p \in M$ and

$$\lambda\colon \tilde{g} \to C^\infty(M;\mathbb{R})$$

is a lift.

Lemma 2.5. Let G be a connected Lie group. If (M,ω,λ) is a Hamiltonian G-space then

$$\lambda(\mathrm{Ad}(g)X) = \lambda(X)\circ g^{-1} .$$

Proof. Let $\phi \in C^\infty(M;\mathbb{R})$, $X \in \mathfrak{g}$. Then

$$\sigma(X)\phi = X_{\lambda(X)}\phi = -\{\lambda(X),\phi\} .$$

Now

$$g_{*p}X^*_p\phi = X^*_p(g^*\phi) = \{\lambda(X),g^*\phi\}$$

$$= -g^*\{(g^{-1})^*\lambda(X),\phi\} .$$

Hence

$$g_{*p}X^*_p = \left(X_{(g^{-1})^*\lambda(X)}\right)_{g\cdot p} .$$

This implies that

$$g_{*g^{-1}p}X^*_{g^{-1}p} = \left(X_{(g^{-1})^*\lambda(X)}\right)_p$$

Now

$$\left(X^*_{g^{-1}\cdot p}f\right) = \frac{d}{dt}\bigg|_{t=0} f(\exp tXg^{-1}\cdot p)$$

$$= \frac{d}{dt}\bigg|_{t=0} f(g^{-1}\exp(t\,\mathrm{Ad}(g)X)\cdot p)$$

$$= g_{*p}^{-1}\sigma(\mathrm{Ad}(g)X)_p\cdot f \quad .$$

Hence

$$\left(X_{(g^{-1})^*\lambda(X)}\right)_p = g_{*_{g^{-1}p}}\, g_{*p}^{-1}\,\sigma(\mathrm{Ad}(g)X)_p$$

$$= \sigma(\mathrm{Ad}(g)X)_p \quad .$$

We have shown

1) $X_{(g^{-1})^*\lambda(X)} = \sigma(\mathrm{Ad}(g)X) \quad .$

This implies that

$(g^{-1})^*\lambda(X) - \lambda(\mathrm{Ad}(g)X) \equiv \mu(g)(X) \in \mathbb{R} \quad ,$

for $g \in G$.

2) $g,X \mapsto \mu(g)(X)$ is a C^∞ map of $G\times g$ to $\mathbb{R}$.

$\mu(g)([X,Y]) = 0$ for all $X,Y \in g$.

The fact that $\mu(g)(X)$ is C^∞ in g,X is obvious. The second statement follows from

$$\begin{aligned}\lambda(\mathrm{Ad}(g)[X,Y]) &= \{\lambda(\mathrm{Ad}(g)X,\lambda(\mathrm{Ad}(g)Y\}\\ &= \{(g^{-1})^*\lambda(X)+\mu(g)(X),(g^{-1})^*\lambda(Y)+\mu(g)(Y)\}\\ &= \{(g^{-1})^*\lambda(X),(g^{-1})^*\lambda(Y)\}\\ &= (g^{-1})^*\{\lambda(X),\lambda(Y)\}\\ &= (g^{-1})^*\lambda([X,Y]) \quad .\end{aligned}$$

We look at μ as a function from G to g^*. Then, if $X,Y \in g$

$$\frac{d}{dt}_{t=0} \mu(g \exp tX)(Y) = \frac{d}{dt}_{t=0} (g \exp tX)^{*-1}\lambda(Y) - \lambda(Ad(g)[X,Y])$$

$$= \frac{d}{dt}_{t=0} \lambda(Y)\circ(\exp -tXg^{-1})-\lambda(Ad(g)[X,Y])$$

$$= -X^*\lambda(Y)\circ g^{-1} - \lambda(Ad(g)[X,Y])$$

$$= g^{*-1}\lambda([X,Y]) - \lambda(Ad(g)[X,Y])$$

$$= \mu(g)([X,Y])$$

$$= 0 \ .$$

Hence $\mu(g)$ is constant. But then

$$\mu(g)(X) = \mu(1)(X) = \lambda(X) - \lambda(X) = 0 \ .$$

Q.E.D.

Now, let (M,ω,λ) be a Hamiltonian G-space. We define

$$\tau: M \to g^*$$

by $\tau(p)(X) = \lambda(X)(p)$. Then

$$\tau(g\cdot p)(X) = \lambda(X)(g\cdot p) = g^*\lambda(X)(p)$$

$$= \lambda(Ad(g^{-1})X)(p) = (g\cdot\tau(p))(X) \ .$$

Hence $\tau: M \to g^*$ is an equivariant C^∞ map. (Equivariant means that $\tau(g\cdot p) = g\cdot\tau(p)$.)

Suppose that M is G-homogeneous. Then, if $p_0 \in M$ is fixed, $\tau(M) = G\cdot\tau(p_0)$. Set

$$f = \tau(p_0) \quad .$$

We have

$$\tau: M \to \mathcal{O}_f$$

is a C^∞, G-equivariant surjective map. We note that

$$\begin{aligned} \tau^*\lambda^f(X)(p) &= \lambda^f(X)(\tau(p)) \\ &= \tau(p)(X) \\ &= \lambda(X)(p) \end{aligned}$$

hence

i) $\tau^*\lambda^f(X) = \lambda(X)$.

By equivariance we have $\tau_*X^* = X^*$ (here the underlying space must be understood from the context). Hence

$$\begin{aligned} (\tau^*\omega^f_{\tau(p)}(X^*,Y^*) &= \omega^f_{\tau(p)}(\tau_*X^*,\tau_*Y^*) \\ &= \{\lambda_f(X),\lambda_f(Y)\}(\tau(p)) \\ &= (\tau^*\{\lambda_f(X),\lambda_f(Y)\})(p) \\ &= (\tau^*\lambda_f([X,Y]))(p) = \lambda([X,Y])(p) \end{aligned}$$

$$= \{\lambda(X),\lambda(Y)\}(p)$$

$$= \omega_p(X^*,Y^*) \quad .$$

This implies that

ii) $\tau^*\omega^f = \omega$.

Proposition 2.5. If (M,ω,λ) is a homogeneous Hamiltonian G-space, then define

$$\tau(p)(X) = \lambda(X)(p) \quad .$$

Fix $p_0 \in M$. Then τ defines a C^∞ map of M onto $\mathcal{O}_f$ satisfying

1) $\tau(g\cdot p) = g\cdot\tau(p)$

2) $\tau^*\lambda_f(X) = \lambda(X), \quad X \in \mathfrak{g}$

3) $\tau^*\omega_f = \omega$

4) τ is a covering map.

Proof. We have proved 1), 2), 3). To prove 4) we note that as manifolds

$$M = G/G_{p_0} \qquad (G_{p_0} = \{g \in G \mid g\cdot p_0 = p_0\}) \; .$$

$$\mathcal{O}_f = G/G_f$$

1) implies

$$\tau(g\cdot p_0) = g\cdot\tau(p_0) = g\cdot f \quad .$$

Hence if $g \in G_{p_0}$, $g \in G_f$. That is,

a) $G_f \supset G_{p_0}$.

The map $\tau\colon M \to \mathcal{O}_f$ is (under the above definitions) just $g\cdot G_{p_0} \to g\cdot G_f$.

b) If H is a Lie group, let H^0 denote its identity component. Then $G_f^0 = G_{p_0}^0$.

Indeed, $\dim M \geq \dim \mathcal{O}_f$. But if $\lambda_f(X) = 0$, then $\lambda(X) = 0$. Thus $\tau_* X^* = 0$ implies X^* (for M) $= 0$. Thus

$$\tau_{*p}\colon T(M)_p \to T(\mathcal{O}_f)_{\tau(p)}$$

is injective. Hence

$$\dim M = \dim \mathcal{O}_f \quad .$$

This implies

$$\dim G_f^0 = \dim G_{p_0}^0 \quad .$$

Hence

$$G_{p_0}^0 = G_f^0 \quad .$$

The proof of b) is therefore complete.

But this implies that

$$\dim G_f = \dim G_{p_0} .$$

Hence, G_f/G_{p_0} is discrete. This says

$$G/G_{p_0} \to G/G_f$$

is a fiber bundle with discrete fiber. Hence, a covering space.

<u>Lemma 2.6</u>. Suppose that $f_1, f_2 \in g^*$ and $\tau: \mathcal{O}_{f_1} \to \mathcal{O}_{f_2}$ satisfies

1) $\tau\cdot(g \cdot p) = g \cdot \tau(p)$

   for $p \in \mathcal{O}_{f_1}$.

2) $\tau^* \lambda_{f_2} = \lambda_{f_1}$

3) $\tau^* \omega^{f_2} = \omega^{f_1}$,

then τ is the identity map.

<u>Proof</u>. All we need is 2). If $p \in \mathcal{O}_{f_2}$, then

$$\lambda_{f_2}(X)(p) = p(X) .$$

Hence

$$\lambda_{f_2}(X)(\tau(p)) = \tau(p)(X)$$

and 2) says

$$\lambda_{f_1}(X)(p) = \lambda_{f_2}(X)(\tau(p))$$

hence

$$p(X) = \tau(p)(X)$$

$$\text{for all } X \in \mathcal{g}, \quad p \in \mathcal{O}_{f_1} .$$

Thus

$$p = \tau(p)$$

$$\text{for all } p \in \mathcal{O}_{f_1} .$$

Q.E.D.

We have proved

Theorem 2.7 (Kostant [16]). If (M,ω,λ) is a homogeneous Hamiltonian G-space, then there is $f \in \mathcal{g}^*$ and an equivariant covering map, $\tau: M \to \mathcal{O}_f$ so that τ is a homomorphism of Hamiltonian G-spaces. (That is, satisfies 1), 2), 3) of Lemma 2.6.) Furthermore, if (M_i,ω_i,λ_i) $i = 1,2$ are homogeneous G-spaces and $\tau_i: M_i \to \mathcal{O}_{f_i}$ is as above, $i = 1,2$, then (M_1,ω_1,λ_1) is locally isomorphic with (M_2,ω_2,λ_2) if and only if $\mathcal{O}_{f_1} = \mathcal{O}_{f_2}$.

Corollary 2.8 (Kostant [16]). If $H^2(\mathcal{g};\mathbb{R}) = 0$ then the most general G-homogeneous symplectic manifold is a G-equivariant covering space of $(\mathcal{O}_f,\omega^f)$ for some $f \in \mathcal{g}^*$.

We note that if g is semi-simple (e.g., G is compact and simply connected), then $H^2(g;\mathbb{R}) = 0$. (Cf. Jacobson [12]).

3. THE CLASSIFICATION OF HOMOGENEOUS SYMPLECTIC MANIFOLDS

Suppose that G is a connected, simply connected Lie group and that (M,ω) is a simply connected homogeneous, symplectic G-space. Then we have seen (Lemma 1.2) that (M,ω) is a strongly symplectic G-space. Arguing as in Section 1 we have seen that there is

$$\mu_0 : g \to C^\infty(M)$$

so that

$$X_{\mu_0(X)} = \sigma(X) \quad .$$

We also saw that

$$\{\mu_0(X),\mu_0(Y)\} - \mu_0([X,Y]) \equiv \beta(X,Y) \in \mathbb{R}$$

and

$$\beta \in C^2(g,\mathbb{R}) , \quad \delta\beta = 0 \quad .$$

If $\beta = \delta\alpha$, $\alpha \in C^1(g,\mathbb{R})$ then we saw that there exists

$$\lambda : g \to C^\infty(M) \quad ,$$

$$\lambda[X,Y] = \{\lambda(X),\lambda(Y)\}$$

such that

$$X_{\lambda(X)} = \sigma(X) .$$

Hence (M,ω) is a homogeneous Hamiltonian G-space. Theorem 2.7 then gives a complete description of M in this case. If $\beta \neq \delta\alpha$, $\alpha \in C^1(g;\mathbb{R})$ then Observation 2.3 applies and (M,ω) is a homogeneous, Hamiltonian $\tilde{G}_\beta$-space.

Let $[\beta]$ denote the class of β in $H^2(g;\mathbb{R})$.

<u>Lemma 3.1</u>. If $\beta,\beta' \in C^2(g;\mathbb{R})$, $\tilde{G}_\beta$ is isomorphic with $\tilde{G}_{\beta'}$ under an isomorphism that is the identity on $\{1\} \times \mathbb{R}$ and $\tilde{G}_\beta/\{1\} \times \mathbb{R}$ if and only if $[\beta] = [\beta']$.

<u>Proof</u>. If $[\beta] = [\beta']$, then $\beta - \beta' = d\alpha$. That is,

$$(\beta - \beta')(X,Y) = -\alpha([X,Y]) .$$

Define

$$A(X,t) = (X, t-\alpha(X)) .$$

Let $[\ ,\ ]_\beta$ and $[\ ]_{\beta'}$ be the bracket on $\tilde{g}_\beta$ and $\tilde{g}_{\beta'}$, respectively. Then

$$A([(X,t),(Y,s)]_\beta) = [A(X,t),A(Y,s)]_{\beta'} .$$

Hence if $[\beta] = [\beta']$, then

$$\tilde{g}_\beta \equiv \tilde{g}_{\beta'} .$$

Suppose $A: \tilde{g}_\beta \to \tilde{g}_{\beta'}$ and $A(0,t) = (0,t)$. Then,

$$A(X,t) = A(X,0) + (0,t) \quad .$$

If

$$A((X,0) + (0,\mathbb{R})) = (X,0) + (0,\mathbb{R})$$

then

$$A(X,0) = (X,\alpha(X)) \quad .$$

Hence,

$$A(X,t) = (X, \alpha(X) + t) \quad .$$

Since A is an isomorphism, this implies

$$\beta(X,Y) = \beta'(X,Y) + \alpha([X,Y])$$

Hence,

$$(\beta - \beta')(X,Y) = \alpha([X,Y]) \quad .$$

Q.E.D.

This discussion implies

Theorem 3.2. If (M,ω) is a simply connected, homogeneous, symplectic G-space, then there exists $\beta \in H^2(g;\mathbb{R})$ and $f \in \tilde{g}^*_\beta$ so that (M,ω) is isomorphic with the universal covering space of $(\mathcal{O}^f,\omega^f)$

4. AN EXAMPLE

Let $M = \mathbb{R}^{2n}$ and

$$\omega = \sum_{i=1}^{n} dx_i \wedge dy_i$$

$(x_1,\ldots,x_n, y_1,\ldots,y_n)$ standard coordinates on $\mathbb{R}^{2n}$ (looked upon as $\mathbb{R}^n \times \mathbb{R}^n$). First of all, the Lie group $\mathbb{R}^{2n} = G$ (the additive group of $\mathbb{R}^{2n}$) acts transitively on $\mathbb{R}^{2n}$ under $x \cdot z = z + x$. Clearly, the action of G on $\mathfrak{g}^*$ is trivial. A basis of $\sigma(\mathfrak{g})$ is

$$\frac{\partial}{\partial x_1}, \ldots, \frac{\partial}{\partial x_n}, \frac{\partial}{\partial y_1}, \ldots, \frac{\partial}{\partial y_n} .$$

$$X_{x_i} = -\frac{\partial}{\partial y_i}, \qquad X_{y_i} = \frac{\partial}{\partial x_i} .$$

Hence if

$$\mu_0\left(\sum a_i \frac{\partial}{\partial x_i} + \sum b_j \frac{\partial}{\partial y_j}\right) = -\sum b_j x_j + \sum a_i y_i$$

then

$$X_{\mu_0(X)} = \sigma(X) .$$

$$\{\mu_0(X), \mu_0(Y)\} - \mu_0([X,Y]) = \beta(X,Y)$$

(here we identify $\mathfrak{g}$ with $\mathbb{R}^{2n}$, $[X,Y] = 0$). Thus

$$\tilde{g}_\beta = \mathbb{R}^{2n} \times \mathbb{R}$$

$$\text{with} \quad [(X,t),(Y,s)] = (0, \beta(X,Y)) \quad .$$

<u>Definition 4.1</u>. $\tilde{g}_\beta$ is called the <u>Heisenberg Lie algebra</u> of dimension $2n+1$ and we will usually denote it h_n.

We now describe $\tilde{G}_\beta$. We give two isomorphic descriptions. The first is, let H_n denote $\mathbb{R}^{2n} \times \mathbb{R}$ with

$$(x,t)\cdot(y,s) = (x+y,\ t+s+\tfrac{1}{2}\beta(x,y)) \quad .$$

It is easily seen that H_n is a Lie group.

<u>Lemma 4.1</u>. H_n is isomorphic with $\tilde{G}_\beta$. H_n is called the <u>Heisenberg group</u> of dimension $2n+1$.

<u>Proof</u>. We first note that if $x_0, t_0 \in H_n$, then $t \mapsto (tx_0, tt_0)$ is a one-parameter subgroup of H_n (this is an easy computation). Thus we can take the Lie algebra of H_n, h, to be $\mathbb{R}^{2n} \times \mathbb{R}$ and $\exp\colon h \to H_n$ to be $\exp(z) = z$. Now,

$$\frac{d}{dt}(tx_0,tt_0)(y,s)(-tx_0,-tt_0)\Big|_{t=0} = [(x_0,t_0),(y,s)]$$

$$= (0,\beta(x_0,y))$$

Hence, h is h_n. Q.E.D.

We give another description of the same Lie group (which is useful in the study of theta functions). Let T_n denote the group of all $(n+2) \times (n+2)$ matrices

$$n(x,y,t) = \begin{bmatrix} 1 & x & t \\ 0 & I_n & y \\ 0 & 0 & 1 \end{bmatrix},$$

x an $1 \times n$ row, y an $n \times 1$ column, I_n the $n \times n$ identity matrix, $t \in \mathbb{R}$. Now T_n has Lie algebra t_n consisting of elements

$$z(x,y,t) = \begin{bmatrix} 0 & x & t \\ 0 & 0 & y \\ 0 & 0 & 0 \end{bmatrix}$$

and

$$[z(x,y,t),z(x',y't')] = z(0,0,\beta((x,y),(x',y')))$$

(here the bracket is $AB - BA$). Hence, $t_n \to h_n$ under $z(x,y,t) \to ((x,y),t)$ is an isomorphism.

We now look for the orbit in h_n^* corresponding to $(\mathbb{R}^{2n},\omega)$. We define (following the discussion in Observation 2.3)

$$\lambda(x,t) = \mu_0(x) + t \quad .$$

Then $(\mathbb{R}^{2n},\omega,\lambda)$ is a Hamiltonian H_n space.

Define (following the discussion preceeding Proposition 2.5) $\tau\colon \mathbb{R}^{2n} \to h_n^*$ by

$$\tau(p)(X,t) = \lambda(X,t)(p) = \mu_0(X)(p) + t \quad .$$

Then

$$\tau(\mathbb{R}^{2n}) = H_n \cdot \tau(0) \subset h_n^* \quad .$$

But

$$\tau(0)(X,t) = t \quad .$$

Set

$$f(X,t) = t, \qquad (X,t) \in h_n \quad .$$

Then we see that

$$H_n \cdot f = \tau(\mathbb{R}^{2n}) \quad .$$

This describes the realization of $(\mathbb{R}^{2n}, \omega)$ as an orbit in h_n^*. We will study this example in more detail when we begin the Kirillov theory.

We look now at a slightly different example. This time, $M = \mathbb{R}^{2n} - \{0\}$, ω is as above. We first look at $\mathbb{R}^{2n}$. Let

$$Sp(n,\mathbb{R}) = G = \{A \in GL(2n,\mathbb{R}) \mid A^*\omega = \omega\} \quad .$$

This is the same as saying that, if

$$J = \begin{bmatrix} 0 & I_n \\ -I_n & 0 \end{bmatrix} \quad ,$$

I_n the $n \times n$ identity matrix, then

$$Sp(n,\mathbb{R}) = \{A \in GL(2n,\mathbb{R}) \mid AJ^tA = J\} .$$

Clearly $(\mathbb{R}^{2n}, \omega)$ is a symplectic G-space. If g is the Lie algebra of G, then g is simple, and hence

$$H^1(g;\mathbb{R}) = H^2(g;\mathbb{R}) = 0$$

(cf. Jacobson [12]). Thus we know from our theory that there is a unique lift $\lambda: g \to C^\infty(\mathbb{R}^{2n})$ so that $X_{\lambda(X)} = \sigma(X)$ and $\lambda([X,Y]) = \{\lambda(X),\lambda(Y)\}$.

If $X \in g$ then X is a $2n \times 2n$ matrix. It is not hard to see that

$$\sigma(X) = \sum_{i,j=1}^{2n} x_{ij} z_j \frac{\partial}{\partial z_i} ,$$

$(z_1,\ldots,z_{2n}) = (x_1,\ldots,x_n, y_1,\ldots,y_n)$. Suppose we wish to find $\phi \in C^\infty(\mathbb{R}^{2n})$ so that $X_\phi = \sigma(X)$. Then

$$X_\phi = \sum_{i=1}^{n} \frac{\partial\phi}{\partial y_i}\frac{\partial}{\partial x_i} - \sum_{i=1}^{n} \frac{\partial\phi}{\partial x_i}\frac{\partial}{\partial y_i} .$$

Hence

$$(*) \qquad \begin{aligned} \frac{\partial\phi}{\partial y_i} &= \sum_{j=1}^{2n} x_{ij} z_j , & i = 1,\ldots,n \\ \frac{\partial\phi}{\partial x_i} &= -\sum_{j=1}^{2n} x_{n+ij} z_j , & i = 1,\ldots,n . \end{aligned}$$

It is natural to look for

$$\phi(z) = \sum a_{ij} z_i z_j \quad ,$$

$$a_{ij} = a_{ji} \quad .$$

If $X \in g$ and if we write X in block form

$$X = \begin{bmatrix} X_1 & X_2 \\ X_3 & X_4 \end{bmatrix}$$

then using $XJ = -J^tX$ we find

a) $X_4 = -{}^tX_1$

b) $X_2 = {}^tX_2$

c) $X_3 = {}^tX_3$

Conversely, if

$$X = \begin{bmatrix} X_1 & X_2 \\ X_3 & X_4 \end{bmatrix}$$

satisfies a), b), c) then $X \in g$. From this and (*) we find

$$a_{ij} = -\frac{1}{2} x_{n+i,j} \quad , \qquad i = 1,\dots,n, \quad j = 1,\dots,2n$$

$$a_{n+ij} = \frac{1}{2} x_{ij} \quad , \qquad i = 1,\dots,n, \quad j = 1,\dots,2n \ .$$

An easy computation gives: If

$$(z,w) = \sum_{i=1}^{2n} z_i w_i$$

then

$$\phi(z) = -\frac{1}{2}(z,JXz) \quad .$$

A direct computation shows that if

$$\lambda(X)(z) = -\frac{1}{2}(z,JXz) \quad ,$$

then

$$\{\lambda(X),\lambda(Y)\} = \lambda([X,Y]) \quad .$$

Since $X_{\lambda(X)} = \sigma(X)$, we have found the unique lift.

Now G acting on $\mathbb{R}^{2n}$ has two orbits, $\{0\}$ and $\mathbb{R}^{2n}-\{0\}$. Hence, $(\mathbb{R}^{2n}-\{0\},\omega,\lambda)$ is a homogeneous, Hamiltonian, $Sp(n,\mathbb{R})$-space. Again, define

$$\tau\colon \mathbb{R}^{2n}-\{0\} \to \mathfrak{g}^*$$

by

$$\tau(p)(X) = \lambda(X)(p) \quad .$$

Then

$$\tau(\mathbb{R}^{2n}-\{0\}) = G\cdot f$$

with, (say),

$$f(X) = \tau(e_1)(X)$$

($e_1,\ldots,e_{2n}$ the standard basis). It is easily seen that

$$f(X) = -\frac{1}{2}(e_{n+1}, Xe_1) .$$

That is, in the jargon of Lie theory, $\{\mathbb{R}^{2n}-\{0\},\omega,\lambda\}$ is the "space of all non-zero long root vectors."

5. WHEN IS $[\omega^f]$ INTEGRAL?

We will need the following result of Palais [20], Theorem III, p. 95.

Theorem 5.1. Let M be a C^∞ manifold. Let $L \subset \mathscr{X}(M)$ be a finite dimensional subalgebra. Suppose that each $X \in L$ generates a global one-parameter group of diffeomorphisms of M. Then there is a connected Lie group G with Lie algebra isomorphic with L so that G acts on M, $\sigma(g) = L$ and $X \mapsto -\sigma(X)$ defines an isomorphism of g with L.

Let G be a connected Lie group. Let (M,ω,λ) be a Hamiltonian G-space. Suppose that $[\omega] \in H^2(M,\mathbb{R})$ is an integral class. We also suppose that $H^1(M,\mathbb{R}) = 0$. Thus there exist L a line bundle over M, ∇ a connection on L and h a Hermitian structure on L that is ∇-invariant. Furthermore, $2\pi i\omega = \text{curv}(L,\nabla)$.

Define

$$\sigma_L: g \to \mathscr{X}(L^\times)$$

by

$$\sigma_L(X) = (\tilde{\delta} \circ \lambda)(X)$$

(See Definition 2.5.7.) Then

$$(\sigma_L(X))^\vee = -X^* .$$

(See Lemma 2.5.4.) Hence, $\sigma_L(X)$ generates a global one-parameter group of diffeomorphisms of $L^\times$. (See Lemma 2.5.2.) Theorem 5.1 says that there is a Lie group, $\tilde{G}$, with Lie algebra isomorphic with $\sigma_L(g)$ so that $\tilde{G}$ acts on $L^\times$ and if $X \in \tilde{g}$ then the map $X \mapsto -\sigma(X) = -X^*$ is an isomorphism of $\tilde{g}$ with $\sigma_L(g)$. Now σ_L is a Lie algebra homomorphism of g with $\sigma_L(g)$. Thus there is a Lie algebra homomorphism of g with $\tilde{g}$. Let $\hat{G}$ be the universal covering group of G. Then there exists

$$\nu: \hat{G} \to \tilde{G}$$

a Lie homomorphism. Let

$$\mu: \hat{G} \to G$$

be the covering homomorphism. Now

$$\sigma_L(g) \subset \underline{e}(L,\nabla)$$

(see the beginning of Section 2.5). Thus, if $g \in \tilde{G}$, then

$$(g\cdot p)\cdot c = g\cdot(p\cdot c)$$

for $p \in L^{\times}$, $c \in \mathbb{C}^{\times}$.

Hence g maps fibers to fibers.

Let G_1 be the image of G in the diffeomorphism group of M. Then, since $\sigma_L(\check{X}) = -X^*$ there is a homomorphism

$$\xi\colon \tilde{G} \to G_1$$

so that

$$p(\tilde{g}\cdot \ell) = \xi(g)p(\ell)$$

for $\ell \in L^{\times}$, $\tilde{g} \in \tilde{G}$.

It is also easy to see that if $x \in M$ then if $\hat{g} \in \hat{G}$,

$$\xi(\nu(g))\cdot x = \mu(g)\cdot x$$

Finally, since $\sigma_L(X) \in \underline{e}(L,\nabla)$ we see that

$$\hat{g}^*\theta = \theta$$

for $\hat{g} \in \hat{G}$.

Here, as usual, θ is the connection form of ∇.

We have proved

Lemma 5.2. Let G be a connected Lie group and let (M,ω,λ) be a Hamiltonian G-space with $H^1(M,\mathbb{R}) = 0$. Let $\hat{G}$ be the universal covering group of G. If $[\omega]$ is integral, then there exists a line bundle L, a connection ∇, and a hermitian structure h on L so that h is ∇-invariant. Furthermore, $\hat{G}$ acts on $L^{\times}$ by bundle homomorphisms $((\hat{g}\cdot\ell)c = \hat{g}\cdot(\ell\cdot c))$ and if $\hat{G} \overset{\mu}{\to} G$ is the covering homomorphism then

$$p(\hat{g}\cdot\ell) = \mu(\hat{g})p(\ell) \quad .$$

Finally,

$$h(\hat{g}\cdot\ell,\hat{g}\cdot s) = h(\ell,s)$$

for $\ell,s \in L^{\times}$.

Only the last line need be proved, but this follows from the fact that if $X \in g$ then

$$X^* \text{ (for } L^{\times}) = -\sigma_L(X)$$

$$= -\tilde{\delta}(\lambda(X)) \in \underline{e}(L,\nabla) \quad .$$

This means, in particular, that

$$X^*|h|^2 = 0 \quad .$$

Now, we have $\hat{G}$ acting on $L^{\times}$ and G acting on M and the actions are compatible relative to $\mu\colon \hat{G} \to G$. Let

$$Z = \operatorname{Ker} \mu \quad .$$

Then Z is a discrete central subgroup of $\hat{G}$.

$$\mu(z)\cdot x = x$$

for all $x \in M$.

Hence, if $z \in Z$, $x \in M$, $\ell \in L_x^\times$, then $z\cdot\ell \in L_x^\times$. Thus

$$z\cdot\ell = \chi_x(z)\cdot\ell$$

for $\ell \in L_x^\times$.

$\chi_x\colon Z \to \mathbb{C}^\times$ is clearly a group homomorphism. Also, since $h(z\cdot\ell, z\cdot\ell) = h(\ell,\ell)$ we have

$$\chi_x\colon Z \to T^1 \subset \mathbb{C}^\times .$$

We therefore see

<u>Observation 5.3</u>. G acts on $L^\times$ compatibly with the action of G on M if and only if for each $x \in M$, $\chi_x(Z) = 1$.

<u>Note</u>. Actually, by noting that the action of B commutes with parallel translations, it is easily seen that χ_x is independent of $x \in M$.

We now assume, in addition, that G acts transitively on M. We fix $p_0 \in M$ and let

$$G_{p_0} = \{g \in G \mid g\cdot p_0 = p_0\} .$$

Let

$$\hat{G}_{p_0} = \mu^{-1}(G_{p_0}) = \{g \in \hat{G} \mid \mu(g)p_0 = p_0\} .$$

Then if $g \in \hat{G}_{p_0}$, $\ell \in L^{\times}_{p_0}$,

$$g \cdot \ell = \chi(g)\ell$$

$$\chi\colon \hat{G}_{p_0} \to \mathbb{C}^{\times}$$

as above (clearly $Z \subset \hat{G}_{p_0}$ so the notation is consistent). Again, since

$$h(g\cdot\ell, g\cdot\ell) = h(\ell,\ell)$$

$$\text{for } g \in \hat{G}$$

we have

$$\chi\colon \hat{G}_{p_0} \to T^1 \subset \mathbb{C}^{\times}$$

is a Lie group homomorphism. Let

$$\chi_*\colon \mathfrak{g}_{p_0} \to i\mathbb{R}$$

be the differential of χ. ($\mathfrak{g}_{p_0}$ is the Lie algebra of $\hat{G}_{p_0}$.)

<u>Lemma 5.4</u>. $\chi_*(X) = 2\pi i\tau(p_0)(X)$.

<u>Proof</u>. Let $X \in \hat{\mathfrak{g}}_{p_0}$. Then

$$\exp(tX)\cdot\ell = e^{t\chi(X)}$$

$$\text{for } \ell \in L_{p_0}.$$

Let S be a cross-section of L, $X \varepsilon g_{p_0}$. If $\ell \varepsilon L_{p_0}$ then

$$\tilde{S}(\exp tX\cdot\ell) = \tilde{S}(e^{t\chi_*(X)}\cdot\ell)$$

$$= e^{-t\chi_*(X)}\tilde{S}(\ell) \quad .$$

Hence

$$X_\ell^*\tilde{S} = -\chi_*(X)\tilde{S}(\ell) \quad .$$

If $X \varepsilon g_{p_0}$ and $\ell \varepsilon L_{p_0}$, then

$$X_\ell^*\tilde{S} = -\eta(\lambda(X), X_{\lambda(X)})_\ell\tilde{S}$$

$$= -\eta(\lambda(X))_\ell\tilde{S} - (\tilde{X}_{\lambda(X)})_\ell\tilde{S} \quad .$$

Now

$$(\tilde{X}_{\lambda(X)})_\ell = 0$$

since $X_{p_0}^* = 0$.

$$\eta(\lambda(X))\tilde{S} = (2\pi i\lambda(X)S)^\sim \quad .$$

Hence

$$X_\ell^*\tilde{S} = -2\pi i\lambda(X)(p_0)\tilde{S}(\ell)$$

for $\ell \varepsilon L_{p_0}$.

Hence

$$\chi_*(X) = 2\pi i\lambda(X)(p_0)$$

as asserted.

We have

<u>Lemma 5.5</u>. Let (M,ω,λ) be a G-homogeneous Hamiltonian G-space. Suppose $H^1(M,\mathbb{R}) = 0$. Fix $p_0 \in M$ and set

$$f = \tau(p_0) \in g^* \quad .$$

A necessary condition for $[\omega] \in H^2(M,\mathbb{R})$ to be integral is that there exist a character χ of $\hat{G}_{p_0}$ so that

$$\chi_*(X) = 2\pi i f(X)$$

for $X \in g_{p_0}$.

Here $\hat{G}$ is the universal covering group of G

$$\hat{G}_{p_0} = \{g \in G | \mu(\hat{g})p_0 = p_0\} \quad ,$$

$\mu: \hat{G} \to G$ the covering homomorphism.

We now prove the converse to this lemma. We may as well assume that G is simply connected. We take $p_0 \in M$, $f = \tau(p_0)$ and assume that there exists a character χ of G_{p_0} so that

$$\chi_*(X) = 2\pi i f(X)$$

for $X \in g_{p_0}$.

We identify M with G/G_{p_0}. We now construct a line bundle over M as follows:

Let G_{p_0} act on the right on $G \times \mathbb{C}$ by

$$(g,z)\cdot h = (gh, \chi(h)^{-1}z) ,$$

$$g \in G, \quad h \in G_{p_0}, \quad z \in \mathbb{C} .$$

Let

$$L = (G \times \mathbb{C})/G_{p_0}$$

relative to this action. Then L defines a line bundle. Here if $[g,z]$ is the orbit of (g,z) in L, then

$$p([g,z]) = gG_{p_0} .$$

(See Wallach [26] for details.) Obviously G acts on L by $g\cdot(X,z) = [gX,z]$.

Define

$$h_{gG_{p_0}}([g,z],[g,w]) = z\overline{w} .$$

Then h is a well defined hermitian structure on L and

$$h(g\ell,gs) = h(\ell,s)$$

$$\text{for } \ell,s \in L, \quad g \in G .$$

We must now define a connection on L. If $S \in C^\infty(M;L)$ and $X \in \mathfrak{g}$ define

$$(\pi(X)S)(y) = \frac{d}{dt}(\exp(-tX)S(\exp tX\cdot y))|_{t=0}$$

We wish to set

$$\nabla_{X_y^*}S = (\pi(X)S)(y) - 2\pi i\lambda(X)(y)S(y)$$

for $S \varepsilon C^\infty(M;L)$, $X \varepsilon \mathfrak{g}$, $y \varepsilon M$.

We must show that $\nabla_{X_y^*}S$ is well defined. That is, if

$$X_y^* = Z_y^*$$

then

$$\nabla_{X_y^*}S = \nabla_{Z_y^*}S \quad .$$

It is enough to check this at $y = p_0$. Then $X - Z \varepsilon \mathfrak{g}_{p_0}$. Set

$$Z = X + U \quad ,$$

$$U \varepsilon \mathfrak{g}_{p_0} \quad .$$

Then

$$(\pi(X+U)S)(p_0) = (\pi(X)S)(p_0) + (\pi(U)S)(p_0) \quad .$$

$$(\pi(U)S)(p_0) = \frac{d}{dt}(\exp(-tU)S)(p_0)|_{t=0}$$

$$= \frac{d}{dt}(e^{2\pi itf(U)}S(p_0))|_{t=0}$$

$$= 2\pi if(U)S(p_0) \quad .$$

Hence

$$(\pi(X+U)S)(p_0) - 2\pi i\lambda(X+U)(p_0)S(p_0)$$

$$= (\pi(X)S)(p_0) + 2\pi i f(U)S(p_0) - 2\pi i\lambda(X)S(p_0) - 2\pi i\lambda(U)(p_0)S(p_0).$$

But

$$\lambda(U)(p_0) = f(U) \quad ,$$

$$U \in g_{p_0} \quad .$$

Hence

$$(\pi(X+U)S)(p_0) - 2\pi i\lambda(X+U)(p_0)S(p_0) = (\pi(X)S)(p_0) - 2\pi i\lambda(X)(p_0)S(p_0)$$

This says that ∇ is well defined.

We also must show that if $S \in C^\infty(M;L)$, $\varphi \in C^\infty(M;\mathbb{C})$, then

$$\nabla_{X_y^*}(\varphi S) = (X_y^*\varphi)S(y) + \varphi(y)\nabla_{X_y^*}S$$

$$\text{for} \quad y \in M, \quad X \in g.$$

But this follows directly from the definitions. Hence, ∇ defines a connection on L.

1) h is ∇-invariant.

Indeed,

$$X^*_y h(S_1,S_2) = \frac{d}{dt}\bigg|_{t=0} h_{\exp tX\cdot y}(S_1(\exp tX\cdot y),S_2(\exp tX\cdot y))$$

$$= \frac{d}{dt}\bigg|_{t=0} h_y(\exp(-tX)S_1(\exp tX\cdot y),\exp(-tX)S_2(\exp tX\cdot y))$$

$$= h_y((\pi(X)S_1)(y),S_2(y)) + h_y(S_1(y),(\pi(X)S_2(y))$$

$$= h_y((\pi(X)S_1)(y) - 2\pi i\lambda(X)(y)S_1(y),S_2(y))$$

$$+ h_y(S_1(y),(\pi(X)S_2)(y) - 2\pi i\lambda(X)(y)S_2(y))$$

$$= h\left(\nabla_{X^*_y}S_1,S_2\right) + h\left(S_1, \nabla_{X^*_y}S_2\right) \quad .$$

As asserted.

2) $\mathrm{Curv}(L,\nabla) = 2\pi i\omega$.

Indeed,

$$[\pi(X),\pi(Y)] = -\pi([X,Y]) \quad .$$

Hence

$$[(\nabla_{X^*} - 2\pi i\lambda(X)),(\nabla_{Y^*} - 2\pi i\lambda(Y))] = -\pi([X,Y]) \quad .$$

Thus

$$[\nabla_{X^*},\nabla_{Y^*}] - 2\pi iX^*\lambda(Y) + 2\pi iY^*\lambda(X) = -\pi([X,Y]) \quad .$$

Now

$$\nabla_{[X^*,Y^*]} = -\nabla_{(X,Y)^*}$$

$$= -\pi([X,Y]) + 2\pi i\lambda([X,Y]) \quad .$$

Therefore,

$$[\nabla_{X^*},\nabla_{Y^*}] - \nabla_{[X^*,Y^*]} = -\pi([X,Y]) + 2\pi i X^*\lambda(Y) - 2\pi i Y^*\lambda(X)$$

$$+ \pi([X,Y]) - 2\pi i\lambda([X,Y])$$

$$= 2\pi i\lambda([X,Y]) \quad .$$

Hence, if $S \in C^\infty(M;L)$,

$$\text{curv}(L,\nabla)(X^*,Y^*)S = 2\pi i\lambda([X,Y])S \quad .$$

But

$$\lambda([X,Y]) = \{\lambda(X),\lambda(Y)\}$$

$$= \omega(X_{\lambda(X)},X_{\lambda(Y)})$$

$$= \omega(X^*,Y^*) \quad .$$

Thus

$$\text{curv}(L,\nabla)(X^*,Y^*) = 2\pi i\omega(X^*,Y^*) \quad .$$

This proves 2).

Combining 1), 2) with Lemma 5.5 we have proved:

Theorem 5.6 (Kostant [16]). Let (M,ω,λ) be a G-homogeneous Hamiltonian G-space. Suppose that $H^1(M;\mathbb{R}) = 0$. Let $\hat{G}$ be the universal covering space of G and let $\mu: \hat{G} \to G$ be the covering homomorphism. Fix $p_0 \in M$. A necessary and sufficient condition for $[\omega]$ to be integral is the existence of a character $\chi: \hat{G}_{p_0} \to T^1$, $\hat{G}_{p_0} = \{g \in G \mid \mu(g)p_0 = p_0\}$ so that $\chi_*(X) = 2\pi i\lambda(X)(p_0)$ for X in the Lie algebra of $\hat{G}_{p_0}$.

Example 5.7. Let G be a connected, simply connected, compact Lie group with Lie algebra $\mathfrak{g}$. Let $\langle\ ,\ \rangle$ be a G-invariant inner product on $\mathfrak{g}$. That is,

$$\langle \mathrm{Ad}(y)X, \mathrm{Ad}(y)Y\rangle = \langle X,Y\rangle$$

$$\text{for } y \in G .$$

Then we can identify $\mathfrak{g}$ with $\mathfrak{g}^*$ under the map $X \to X^\#$, $X^\#(Y) = \langle X,Y\rangle$.

Let $T \subset G$ be a maximal torus with Lie algebra h_* $\mathfrak{g}$. Then

$$\mathrm{Ad}(G)\cdot h_* = \mathfrak{g} .$$

Let $W(T)$ be the Weyl group of T that is

$$W(T) = N(T)/T ,$$

$N(T)$ the normalizer of T in G. Then $N(T)/T$ is a finite group. Furthermore, if $h_1, h_2 \in h_*$

$$\mathrm{Ad}(G)h_1 \cap \mathrm{Ad}(G)h_2 \neq \emptyset$$

then there is $g \in N(T)$ so that $gh_1 = h_2$. Clearly $W(T)$ acts on h_* by $S \cdot h = g \cdot h$ for $g \in S$.

We therefore see that the orbit space $g/\mathrm{Ad}(G) = h_*/W$. Let $h_0 \in h_*$. Then set

$$G_{h_0} = \{g \in G \mid \mathrm{Ad}(y) \cdot h_0 = h_0\} .$$

1) G_{h_0} is connected. (Cf. Wallach [26].)

1) implies G/G_{h_0} is simply connected. Let ω^h be the corresponding simplectic form on

$$G/G_{h_0} = G \cdot h_0 = \mathcal{O}^{h_0} .$$

Let us see what Theorem 6.3 says in this case. Let

$$f(X) = \langle h_0, X\rangle .$$

Then ω^h is integral if and only if

$$X \mapsto 2\pi i f(X) ,$$

$$X \in g_{h_0} .$$

is the differential of a character of G_u. (Here

$$g_{h_0} = \{X \in g \mid [X, h_0] = 0\} .)$$

Now $T \subset G_{h_0}$. Thus, in particular, we see that

$$\chi(\exp X) = e^{2\pi i f(X)}$$

defines a character of T. Now $\exp\colon h_* \to T$ is a covering homomorphism (looking at h_* as a Lie group under addition). Let $\Gamma \subset h_*$ be the kernel of exp. Then, if ω^{h_0} is integral $f(\Gamma) \subset \mathbb{Z}$.

We recall the description of Γ. Let $g_{\mathbb{C}}$ be the complexification of g. Then under $\mathrm{Ad}(T)$,

$$g_{\mathbb{C}} = h + \sum_{\alpha\in\Delta} g_\alpha$$

where h is the complexification of h_* and $\Delta \subset h^*$,

$$g_\alpha = \{X \in g_{\mathbb{C}} \mid \mathrm{Ad}(H)X = \alpha(H)X \text{ for } H \in h\} .$$

Furthermore, defining

$$\xi_\alpha(\exp H) = e^{\alpha(H)} ,$$

ξ_α defines a character of T and if $t \in T$, $X \in g_\alpha$

$$\mathrm{Ad}(t)X = \xi_\alpha(t)X .$$

Clearly, this implies, in particular, that

$$\Delta \subset i(h_*)^* .$$

We extend $<\ ,\ >$ to a complex bilinear form on $g_{\mathbb{C}}$. Then

$\langle\ ,\ \rangle$ on ih_* is negative definite. If $\alpha \in \Delta$ let $h_\alpha \in ih_*$ be such that

$$\langle h_\alpha, H\rangle = \alpha(H) ,$$
$$H \in h .$$

Then Γ is the lattice in h_* spanned over $\mathbb{Z}$ by the elements

$$2\pi i\left(\frac{2}{\langle h_\alpha, h_\alpha\rangle}\ h_\alpha\right) ,$$
$$\alpha \in \Delta$$

(For a proof, see Wallach, [26], Chapter 4.)

Thus if $f(\Gamma) \subset \mathbb{Z}$ then we must have

$$2\pi\ \frac{2}{\langle h_\alpha, h_\alpha\rangle}\ \langle ih_\alpha, h_0\rangle \in \mathbb{Z}\ \cdot$$
$$\alpha \in \Delta$$

Let

$$\lambda = 2\pi i f .$$

Then this says

$$(*) \qquad 2\,\frac{\lambda(h_\alpha)}{\alpha(h_\alpha)} \in \mathbb{Z}$$
$$\text{for } \alpha \in \Delta$$

Now using the representation theory of compact groups (cf. Wallach, [26]) we have:

Lemma 5.8. $[\omega^{h_0}]$ is integral if and only if $\lambda(H) = 2\pi i\langle h_0, H\rangle$ satisfies (*) above.

This says that $[\omega^{h_0}]$ is integral if and only if $2\pi i f$ is integral in the sense of compact Lie groups. Under this condition there is a line bundle L over θ^{h_0}, hermitian structure h on L and a connection ∇ on L so that h is ∇-invariant.

Finally,

$$\operatorname{curv}(L,\nabla) = 2\pi i\omega^{h_0} .$$

Example 5.9. This example is a special case of Example 5.7. Let

$$G = SU(n) , \qquad n \geq 2 .$$

$$\mathfrak{g} = \{X \in M_n(\mathbb{C}) \mid {}^t\overline{X} = -X, \operatorname{tr} X = 0\} .$$

Take T to be the diagonal elements of G.

$$\mathfrak{h}_* = \{\operatorname{diag}(i\lambda_1,\ldots,i\lambda_n) \mid \lambda_i \in \mathbb{R}, \Sigma\lambda_i = 0\} .$$

$$\exp(\operatorname{diag}(i\lambda_1,\ldots,i\lambda_n)) = \operatorname{diag}(e^{i\lambda_1},\ldots,e^{i\lambda_n}) .$$

Thus

$$\Gamma = 2\pi i\cdot\{\operatorname{diag}(\lambda_1,\ldots,\lambda_n) \mid \Sigma\lambda_i = 0, \lambda_i \in \mathbb{Z}\} .$$

Thus if $h_0 \in \mathfrak{h}_*$ and ω^h is integral and if $\langle X,Y\rangle = -\operatorname{tr} XY$,

$X, Y \in g$ then $[\omega^{h_0}]$ is integral if and only if $\langle h_0, \Gamma \rangle \subset \mathbb{Z}$. Thus if

$$h_0 = i \operatorname{diag}(\lambda_1, \ldots, \lambda_n)$$

we must have

$$2\pi \sum k_i \lambda_i \in \mathbb{Z}$$

$$\text{for all } k_i \in \mathbb{Z}, \quad \sum k_i = 0 \quad .$$

In particular, this says that

$$2\pi(\lambda_i - \lambda_j) \in \mathbb{Z}$$

$$\text{for } i \neq j \ .$$

In particular, this says that

$$2\pi\lambda_{n-1} = 2\pi\lambda_n + m_{n-1},\ 2\pi\lambda_{n-2} = 2\pi\lambda_{n-1} + m_{n-2}, \ldots, 2\pi\lambda_1 = 2\pi\lambda_2 + m_1,$$

$m_i \in \mathbb{Z}$.

Since

$$\sum \lambda_i = 0$$

this says

$$2\pi n\lambda_n + \sum_{j=1}^{n-1} rj \cdot m_j = 0 \quad .$$

We therefore see that $[\omega^{h_0}]$ is integral if and only if there are integers $m_1,\ldots,m_{n-1}$ so that

$$h_0 = \frac{i}{2\pi}\operatorname{diag}(m_1+\cdots+m_{n-1}, m_2+\cdots+m_{n-1},\ldots,m_{n-1},0)$$

$$- \frac{i}{2\pi n}\left(\sum_{j=1}^{n-1} jm_j\right) I .$$

<u>Example 5.10</u>. $P^n(\mathbb{C})$. This is a special case of Example 5.9. We look at $SU(n+1)$. Let $k \in \mathbb{Z}$ and

$$h_k = \frac{k}{2\pi i}\operatorname{diag}(1,\ldots,1,0) - \frac{kn}{(n+1)2\pi i} I .$$

Then ω^{h_k} is integral and

$$G_{h_k} = \left\{ \left[\begin{array}{c|c} g & 0 \\ \hline 0 & (\det g)^{-1} \end{array}\right] \;\middle|\; g \in U(n) \right\} .$$

Thus for $k \in \mathbb{Z}$ we have

$$L_k \to SU(n+1)/U(n) = P^n(\mathbb{C})$$

a line bundle with chern class given by ω^{h_k}.

Now,

$$\omega^{h_k}(X^*,Y^*) = -\operatorname{tr} h_k[X,Y] = -k \operatorname{tr} h_1[X,Y]$$

$$= k\omega^{h_1} .$$

Set

$$\omega^{h_1} = \omega \quad .$$

We look for

$$L_1 \to P^n(\mathbb{C})$$

the line bundle constructed as above for ω. Let

$$\mathbb{L} = \{(p,v) \mid v \in p,\ p \in P^n(\mathbb{C})\} \quad .$$

Let $SU(n+1)$ act on $P^n(\mathbb{C})$ by

$$g\cdot[z] = [g\cdot z] \quad ,$$

$$z \in \mathbb{C}^{n+1} - \{0\}$$

Then $g\cdot(p,v) = (g\cdot p, g\cdot v)$ defines an action of G on $\mathbb{L}$ so that

$$g\cdot\mathbb{L}_x = \mathbb{L}_{g\cdot x} \quad .$$

Let

$$p_0 = [0,\ldots,0,1] \quad .$$

Then

$$G_{p_0} = U(n)$$

$$= \left\{ \left[\begin{array}{c|c} g & 0 \\ \hline 0 & (\det g)^{-1} \end{array} \right] \middle| \ g \in U(n) \right\} \quad .$$

If $f \in G_{p_0}$, $v \in \mathbb{L}_{p_0}$, then

$$v = (0,\ldots,0,z)$$

$$z \in \mathbb{C} .$$

Hence

$$g\cdot(p_0,v) = (p_0, \det(g)^{-1}z) .$$

Thus in the construction above

$$\chi \left[\begin{array}{c|c} g & 0 \\ \hline 0 & (\det g)^{-1} \end{array}\right] = \det g^{-1} .$$

If $X \in g_{p_0}$ then $\chi_*(X) = 2\pi i\langle h_1,X\rangle$ by the obvious computation. Also, it is clear from the above construction that

$$\begin{aligned} \mathbb{L} &= L_{p_0}\, i(K(\mathbb{L})) \\ &= d_1 R\cdot[\omega^{h_1}] . \end{aligned}$$

APPENDIX TO CHAPTER 3

TORI

Let V be a 2n-dimensional vector space over $\mathbb{R}$. Let $\Gamma \subset V$ be a discrete subgroup of V (V is a Lie group under addition) so that $T = V/\Gamma$ is compact. Then T is a (2n) torus. T is clearly a Lie group. We write its group law multiplicatively. Let $\pi\colon V \to T$ be the natural map. Then $\pi(x+y) = \pi(x)\pi(y)$.

Suppose now that ω is a symplectic structure on T so that $L_t^*\omega = \omega$ for $t \in T$. (Here $L_t x = tx$.)

Then $\pi^*\omega \in \mathcal{D}^2(V;\mathbb{R})$ is translation invariant. We identify V with $T(V)_x$ at each $x \in V$ be taking

$$X(v)_x f = \frac{d}{dt} f(x + tv)\Big|_{t=0}$$

$$\text{for } v \in V .$$

Then if $T(x)y = y + x$ clearly $T(x)_{*y}X(v)_x = X(v)_{x+y}$. Hence

$$(\pi^*\omega)_x(X(v), X(\omega)) = \beta(v,\omega)$$

with $\beta\colon V \times V$ a non-degenerate alternating bilinear form.

<u>Lemma A3.1</u>. $[\omega] \in H^2(T;\mathbb{R})$ is integral only if $\beta(\Gamma,\Gamma) \subset \mathbb{Z}$.

<u>Proof</u>. Let $\gamma,\delta \in \Gamma$ be such that $\beta(\gamma,\delta) \neq 0$. Let

$$V_1 = \mathbb{R}\gamma + \mathbb{R}\delta .$$

Then $\dim V_1 = 2$. Let $\Gamma_1 = V_1 \cap \Gamma$. Then Γ_1 is a discrete subgroup of V_1 and $V_1/\Gamma_1 = T_1$ is a torus. Let $f: V_1 \to T$ be defined by $f(v_1) = \pi(v_1)$. Then

$$f(v_1 + \xi) = f(v_1)$$

$$\text{for} \quad \xi \in \Gamma_1 \quad .$$

Hence f induces a C^∞ map of T_1 into T,

$$\varphi(v_1 + \Gamma_1) = f(v_1) \quad ,$$

$$v_1 \in V_1 \quad .$$

Clearly f is an imbedding of T_1 into T. Now from our constructions of Chapter 1 it is easy to see that if ω is integral than $\varphi^*\omega$ is integral for T_1. Let

$$\beta_1 = \beta|_{V_1 \times V_1} \quad .$$

Then if $\pi_1: V_1 \to T_1$ is the canonical map,

$$\pi_1^*\varphi^*\omega = (X(v), X(\omega)) = \beta_1(v,\omega) \quad .$$

We note that $H^2(T_1, \mathbb{Z}) = \mathbb{Z}$, hence $\varphi^*\omega$ is integral if and only if

$$\int_{T_1} \varphi^*\omega \in \mathbb{Z} \quad .$$

But if γ_1, γ_2 is a basis for Γ_1, then

$$\int_{T_1} \varphi^*\omega = \beta_1(\gamma_1,\gamma_2) \quad .$$

Hence, since $\gamma = m_1\gamma_1 + m_2\gamma_2$,

$$\delta = n_1\gamma_1 + n_2\gamma_2 \quad , \quad \beta(\gamma,\delta) \in \mathbb{Z} \quad .$$

Actually, the condition of Lemma A.5.1 is necessary and sufficient. We will see this in the course of our discussion.

Our previous theory tells us that if we wish to find a line bundle for the translation invariant symplectic structure $\pi^*\omega$ on V it is necessary to use a certain extension of V. A perusal of the results of Chapter 3 tells us that the extension to study is

$$h_\beta = V \underset{\beta}{\times} \mathbb{R} = V \times \mathbb{R}$$

with Lie algebra structure

$$[(v,t),(\omega,s)] = [0, \beta(v,\omega)] \quad .$$

We now find the simply connected Lie group corresponding to h_β. The Cambell-Hausdorf formula suggests the following group law: H_β should be $V \times \mathbb{R}$ with multiplication

$$(*) \qquad (v,t)(\omega,s) = (v+\omega,\ t+s+\tfrac{1}{2}\beta(v,\omega))$$

(See Section 5, Chapter 3.) It is easy to check that (*) does indeed define a Lie group structure on $V \times \mathbb{R}$. The Lie algebra of $V \times \mathbb{R}$ relative to (*) can be identified with $V \times \mathbb{R}$. If $(v,u) \in V \times \mathbb{R}$, then (*) says $t \to (tv,tu)$ is a one-parameter group. Thus $\exp(v,u) = (v,u)$ under all of these identifications.

Now

$$\exp(tv,ts)\,\exp(\omega,u)\,\exp(-tv,-ts) = \exp(e^{t\,\mathrm{ad}(v,s)}(\omega,u)).$$

But then

$$e^{t\,\mathrm{ad}(v,s)}(\omega,u) = (\omega,\ u+t\,(v,\omega)) .$$

Thus

$$\mathrm{ad}(v,s)\cdot(\omega,u) = (0,\ \beta(v,\omega)) .$$

This implies that if we take H_β to be $V \times \mathbb{R}$ with group law given by (*), then H_β does indeed have Lie algebra h_β.

Let

$$z = \{(0,t) \mid t \in \mathbb{R}\} .$$

Then z is the center of H_β. Clearly the map

$$V \xrightarrow{\psi} H_\beta/z \quad , \quad v \xrightarrow{\psi} (v,0)z$$

is a diffeomorphism of V with H_β/z. Let

$$\pi\colon H_\beta \to H_\beta/z$$

be the natural map. Set

$$H_\beta(\Gamma) = \pi^{-1}(\psi(\Gamma)) .$$

Then clearly

$$H_\beta(\Gamma) = \{(\gamma,t) \mid \gamma \in \Gamma,\ t \in \mathbb{R}\}$$

is a normal subgroup of H_β. Let $\phi\colon V \to H_\beta/H_\beta(\Gamma)$ be given by $\phi(v) = (v,0)H_\beta(\Gamma)$. Then

$$\begin{aligned}\phi(v+\gamma,0) &= (v+\gamma,0)H_\beta(\Gamma) \\ &= (v,0)\cdot(\gamma,\ \tfrac{1}{2}\beta(\gamma,v))H_\beta(\Gamma) \\ &= (v,0)H_\beta(\Gamma) \quad .\end{aligned}$$

Hence ϕ induces a map $\Phi\colon T \to H_\beta/H_\beta(\Gamma)$. It is easy to see that Φ is a diffeomorphism.

Now suppose that ω is a symplectic structure on T that is left invariant on T. Let $\beta\colon V \times V \to \mathbb{R}$ be as above and suppose that $\beta(\Gamma,\Gamma) \subset \mathbb{Z}$.

We define a line bundle over T in the usual manner. Suppose that $\chi\colon H_\beta(\Gamma) \to T^1$ is a homomorphism (continuous). We suppose for simplicity that $\chi((0,t)) = \exp(2\pi it)$. (Actually, it is an easy exercise to see that if $\chi\colon H_\beta(\Gamma) \to T^1$ is a continuous homomorphism, then $\chi(0,t) = \exp(2\pi ikt)$ with $k \in \mathbb{Z}$.) We construct a line bundle over T as follows: Let $H_\beta(\Gamma)$ act on $H_\beta \times \mathbb{C}$ by $(g,z)h = (gh,\chi(h)^{-1}z)$. Then

$$L_\chi = H_\beta \times \mathbb{C}/H_\beta$$

under this action.

Of course, before we go on we must show that there is such a χ. We will actually describe all such χ. We will use the following lemma.

<u>Lemma A.3.2</u>. There is a basis $e_1,\ldots,e_{2n}$ for Γ over $\mathbb{Z}$ so that

$$\beta(e_i,e_j) = \beta(e_{n+i},e_{n+j}) = 0$$

$$\text{for } i,j \leq n$$

and

$$\beta(e_i,e_{n+i}) = m_i ,$$

$$i = 1,\ldots,n, \quad m_i \in \mathbb{Z}, \quad m_i > 0$$

and

$$m_1|m_2| \cdots |m_n .$$

Furthermore, $m_1\cdots m_n$ is independent of the choice of basis of Γ as above.

<u>Proof</u>. The independence assertion is easy since

$$m_1\cdots m_n = \Omega(e_1,\ldots,e_{2n})$$

where Ω is the Pfaffian of β. If $f_1,\ldots,f_{2n}$ is another basis for Γ over $\mathbb{Z}$, then

$$f_i = \sum_{j=1}^{2n} a_{ji}e_j ,$$

$$a_{ji} \in \mathbb{Z}, \quad e_i = \sum_{j=1}^{2n} a^{ji}f_j, \text{ and } a^{ij} \in \mathbb{Z} .$$

Thus, if $A = (a_{ij})$, $A^{-1} = (a^{ij})$ and thus $\det A \in \mathbb{Z}$ and $(\det A)^{-1} \in \mathbb{Z}$. Thus $\det A = \pm 1$. This implies that

$$m_1 \cdots m_n = |\Omega(f_1,\ldots,f_{2n})| .$$

The existence of the $\mathbb{Z}$ basis is also easy. Let

$$I = \beta(\Gamma,\Gamma) \subset \mathbb{Z} .$$

Then it is clear that I is an ideal is $\mathbb{Z}$. Hence

$$I = Zm_1$$

$$m_1 \in \mathbb{Z}, \quad m_1 > 0 .$$

Let $e_1, e_{n+1} \in \Gamma$ be such that $\beta(e_1,e_{n+1}) = m_1$. Let

$$\Gamma_1 = \{r \in \Gamma |\ \beta(e_1,\gamma) = (e_{n+1},\gamma) = 0\}$$

$$V_1 = \{v \in V |\ \beta(e_1,v) = \beta(e_{n+1},v) = 0\} .$$

We assert that

$$\Gamma = \mathbb{Z}e_1 + \mathbb{Z}e_{n+1} + \Gamma_1 .$$

In fact, if $\xi \in \Gamma$ set

$$\xi' = + \frac{1}{m_1} \beta(e_1,\xi)e_{n+1} - \frac{1}{m_1} \beta(e_{n+1},\xi)e_1 .$$

Then

$$\beta(e_1, \xi-\xi') = \beta(e_1,\xi) - \frac{1}{m_1} \beta(e_1,\xi)\beta(e_1, e_{n+1})$$

$$= 0 .$$

Similarly,

$$\beta(e_2, \xi-\xi') = 0 .$$

Hence

$$\xi - \xi' \in \Gamma_1 .$$

Continue the proof with the obvious induction. Q.E.D.

If $v \in V$, $v = \sum v_i e_i$ (e_i as in A.3.2), define

$$F(v) = \sum_{i=1}^{n} v_i v_{n+1} m_i .$$

Then we assert that

$$(*) \qquad \chi((\gamma,t)) = e^{2\pi it} e^{\pi i F(\gamma)}$$

is a desired character. This can be checked by the obvious computation. We note that any $F\colon \Gamma \to \mathbb{R}$ such that

$$F(\gamma) + F(\delta) + \beta(\gamma,\delta) - F(\gamma+\delta) \in 2\mathbb{Z}$$

will also define a character by the above formula. If χ' is another such character, then $\chi^{-1}\chi'(0,t) \equiv 1$. Hence,

$$\chi^{-1}\chi'\left(\sum n_i e_i, t\right) = e^{2\pi i \Sigma \ell_i n_i}$$

with $(\ell_1,\ldots,\ell_{2n}) \in \mathbb{R}^{2n}$.

We fix one such χ and we have

$$L \xrightarrow{p} T$$

a line bundle with H_β acting on L_χ by bundle maps compatible with the action of H_β on $T = H_\beta/H_\beta(\Gamma)$.

We define $h((g,z),(g,w)) = z\overline{w}$ as usual. Let

$$(\rho(X)s)(z) = \frac{d}{dt} \exp(-tX)s(\exp tX\cdot z)|_{t=0}$$

for $X \in h_\beta$, $s \in C^\infty(T;L_\chi)$.

Again by working by analogy with the case of line bundles over V we define a connection on L_χ by

$$(**) \qquad \nabla_{X_z^*} s = (\rho(X)s)(z) + 2\pi its(z)$$

for $X = (x,t) \in h_\beta$, $s \in C^\infty(T;L_\chi)$

To see that $\nabla_{X_z^*} s$ is well defined we must show that if $X_z^* = 0$, $\nabla_{X_z^*} s = 0$. But if $X_z^* = 0$, then

$$X = (0,U)$$

for some $U \in \mathbb{R}$.

$$\exp(tX)\cdot z = z$$

for all $t \in \mathbb{R}$, $z \in T$.

Hence

$$\exp(-tX)s(\exp tX\cdot z) = \exp(-tX)s(z) \quad .$$

Now

$$s(z) = [g,w]$$

for some $g \in H$, $w \in \mathbb{R}$.

$$\begin{aligned} \exp(-tX)\cdot[g,w] &= (0,-tX)\cdot[g,w] \\ &= [(0,-tu)g,w] \\ &= [g(0,-tu),w] \\ &= e^{-2\pi itu}[g,w] \quad . \end{aligned}$$

Thus

$$\varphi(X)s(z) = -2\pi ius(z) \quad .$$

Hence (**) is well defined. It is now clear that (**) defines a connection ∇ on L_X and that h is ∇-invariant.

A direct computation of $[\nabla_{X^*},\nabla_{Y^*}] - \nabla_{[X^*,Y^*]}$ gives

<u>Lemma A.3.3</u>. Let L_X,∇ be defined as above for as above, then

$$\text{curv}(L_X,\nabla) = 2\pi i\omega \quad .$$

<u>Note</u>. A.3.3 gives the promised converse to Lemma A.3.1. Of course, we have proved much more.

Suppose now that $\omega \in \mathcal{D}^2(T,\mathbb{R})$ is an arbitrary symplectic structure. We note that since T is connected $L_t^*\omega$ and ω give the same de Rham cohomology class. Using the singular cohomology version of de Rham's theorem it is easy to see that

$$\int_T (L_t^*\omega)\, dt = \omega^0$$

($\int_T dt = 1$, dt invariant measure on T), is in the same cohomology class as ω (see the discussion in Wallach [26], 3.8).

But then the Pfaffian of ω^0 must be in the same class of that of ω. Hence,

$$\underbrace{\omega^0 \wedge \cdots \wedge \omega^0}_{\text{n-times}} \neq 0 \quad .$$

Thus, ω^0 is non-degenerate and left invariant.

Let now (L,∇) be a pair of line bundles L, and a connection on L so that

$$\text{curv}(L,\nabla) = 2\pi i\omega \quad .$$

Now $\omega - \omega^0 = -d\eta$, $\eta \in \mathcal{D}^1(T,\mathbb{R})$. We define a new connection $\tilde{\nabla}$ on L as follows:

$$\tilde{\nabla}_v s = \nabla_v s + 2\pi i\eta(v)s(v) \quad .$$

Then it is clear that $\tilde{\nabla}$ is a connection on L and

$$\tilde{\nabla}_X\tilde{\nabla}_Y s - \tilde{\nabla}_Y\tilde{\nabla}_X s - \tilde{\nabla}_{[X,Y]}s$$

$$= \tilde{\nabla}_X(\nabla_Y s = 2\pi i\eta(Y)s) - \tilde{\nabla}_Y(\nabla_X s + 2\pi i\eta(X)s) - \nabla_{[X,Y]}s - 2\pi i\eta(X,Y)s$$

$$= \nabla_Y\nabla_X s + 2\pi i(X\eta(Y))s + 2\pi i\eta(Y)\nabla_X s + 2\pi i\eta(X)\nabla_Y s - (4\pi)^2\eta(X)\eta(Y)s$$

$$- \nabla_Y\nabla_X s - (2\pi i)Y\eta(X)s - 2\pi i\eta(X)\nabla_Y s - 2\pi i\eta(Y)\nabla_X s$$

$$+ (4\pi)^2\eta(Y)\eta(X)s + \nabla_{[X,Y]}s - 2\pi i\eta(X,Y)s$$

$$= 2\pi i\omega(X,Y)s + 2\pi i(X\eta(Y) - Y\eta(X) - \eta([X,Y])s$$

$$= 2\pi i\omega^0(X,Y)s \quad .$$

Suppose now that V is actually an n-dimensional vector space over $\mathbb{C}$. Then, still looking at V as a real vector space, we then see that multiplication by i gives a real linear map $J: V \to V$, $J^2 = -I$. Γ acting by translations on V preserves the complex structure. Hence, V/Γ is an n-dimensional complex manifold.

We say that T is an abelian variety if there exists a Hermitian structure $< , >$ on T (relative to the complex structure J) so that if $\omega(X,Y) = \text{Im} <X,Y>$. Then

1) $d\omega = 0$

2) ω is integral.

Again, we may take $< , >$ and modify it by taking

$$(v,\omega) = \int_T <t_* v, t_* \omega> dt \quad .$$

Then $\text{Im}(X,Y) = \omega^0(X,Y)$ as above.

Let us start with ω^0. Then, as above, we have $\beta: V \times V \to \mathbb{R}$ non-degenerate and alternating, $\beta(\Gamma,\Gamma) \subset \mathbb{Z}$. Let χ be a character of $H_\beta(\Gamma)$. We now show that $L_\chi \to T$ is actually a holomorphic line bundle.

To do this we will need a special case of what is called a complex polarization (see [1]).

Let

$$V_{\mathbb{C}} = V \otimes_{\mathbb{R}} \mathbb{C} .$$

Let the complex structure on V be denoted J. Let

$$V^{+} = \{v \in V_{\mathbb{C}} \mid Jv = iv\} ,$$

$$V^{-} = \{v \in V_{\mathbb{C}} \mid Jv = -iv\} .$$

Extend β to $V_{\mathbb{C}} \times V_{\mathbb{C}}$ to be complex bilinear.

a) $\beta|_{V^+ \times V^+} = \beta|_{V^- \times V^-} = 0$.

Indeed, if $w, v \in V^{+}$,

$$i\beta(v,w) = \beta(Jv,w) = \beta(JJv,Jw)$$

$$= -\beta(v,Jw) = i\beta(v,w)$$

(Here we note that $\beta(Jv,Jw) = \beta(v,w)$ since β' on $V \times V$ is just $\operatorname{Im} \langle\ ,\ \rangle$ and $\operatorname{Im} \langle iv,iw\rangle = \operatorname{Im} \langle v,w\rangle$.)

Let $H^{\mathbb{C}} = V_{\mathbb{C}} \times \mathbb{C}$ with multiplication

$$(v,z)(w,u) = (v+w,\ z+u+\tfrac{1}{2}\beta(v,w)) .$$

We look at V as a <u>real</u> subspace of $V_{\mathbb{C}}$. Let

$$H_{\beta}^{\mathbb{C}}(\Gamma) = \{(\gamma,z) \mid \gamma \in \Gamma,\ z \in \mathbb{C}\} .$$

We extend χ to a holomorphic character on $H_\beta^{\mathbb{C}}(\Gamma)$ by setting $\chi(\gamma,z) = \chi(\gamma,0)e^{2\pi iz}$.

Let $\tilde{L}_\chi$ be the line bundle over $H_\beta^{\mathbb{C}}/H_\beta^{\mathbb{C}}(\Gamma)$ given as usual by $H_\beta^{\mathbb{C}} \times \mathbb{C}/H_\beta^{\mathbb{C}}(\Gamma)$ with $(g,z)\cdot h = (gh,\chi(h)^{-1}z)$. Then

$$\tilde{L}_\chi \to H_\beta^{\mathbb{C}}/H_\beta^{\mathbb{C}}(\Gamma)$$

is a holomorphic line bundle where $H_\beta^{\mathbb{C}}/H_\beta^{\mathbb{C}}(\Gamma)$ is given the natural complex structure since $H_\beta^{\mathbb{C}}(\Gamma)$ is a closed complex subgroup of $H_\beta^{\mathbb{C}}$.

Let

$$N^- = \{(v,0) \mid v \in V^-\} \quad .$$

Then $H_\beta^{\mathbb{C}}$ acts on the left on $\tilde{L}_\chi$ and $H_\beta^{\mathbb{C}}/H_\beta^{\mathbb{C}}(\Gamma)$ by consistent holomorphic automorphisms. It is an exercise to see that N^- acts freely and properly discontinuously on both. We can thus form

$$\hat{L}_\chi = N^-\backslash\tilde{L}_\chi \to N^-\backslash H_\beta^{\mathbb{C}}/H_\beta^{\mathbb{C}}(\Gamma) \quad .$$

It is easy to check that

$$N^-\backslash H_\beta^{\mathbb{C}}/H_\beta^{\mathbb{C}}(\Gamma) = V^+/P^+(\Gamma)$$

where

$$P^+(v) = \frac{1}{2}(v - iJv) \quad ,$$

$$P^-(v) = \frac{1}{2}(v + iJv)$$

for $v \in V_{\mathbb{C}}$.

We assert that $\hat{L}_\chi \to V^+/P^+(\Gamma)$ is isomorphic with $L_\chi \to V/\Gamma$. (Note that $P^+: V \to V^+$ satisfies $P^+(Jv) = iP^+(v)$ thus, the map

$$V/\Gamma \xrightarrow{h} V^+/P^+(\Gamma)$$

given by $h(v+\Gamma) = P^+(v) + P^+(\Gamma)$ is a holomorphic diffeomorphism.) To see this we note that if

$$\tilde{Z} = \{(0,it) \mid t \in \mathbb{R}\} .$$

Then, $H_\beta^{\mathbb{C}} = N^- H_\beta \tilde{Z}$, $H_\beta^{\mathbb{C}}(\Gamma) = H_\beta(\Gamma)\tilde{Z}$. Thus we may define

$$\mu\colon H_\beta/H_\beta(\Gamma) \to N^-\backslash H_\beta^{\mathbb{C}}/H_\beta^{\mathbb{C}}(\Gamma)$$

by

$$hH_\beta(\Gamma) \to N^- hH_\beta(\Gamma)$$

and

$$\nu\colon L_\chi \to L_\chi$$

by

$$\nu([h,u]) = N^-[h,u] .$$

Then ν defines a C^∞ bundle mpa of $L_\chi \to \hat{L}_\chi$ so that $p(\nu(\ell)) = \mu(p(\ell))$. Hence we have given L_χ the structure of a holomorphic line bundle.

Finally we describe the holomorphic cross-sections of L_χ. If f is a holomorphic cross-section of $\hat{L}_\chi$ then

$$f\colon N^-\backslash H_\beta^{\mathbb{C}}/H_\beta^{\mathbb{C}}(\Gamma) \to \hat{L}_\chi$$

is holomorphic and $pf(z) = p(z)$. Let

$$q\colon H_\beta^{\mathbb{C}}/H_\beta^{\mathbb{C}}(\Gamma) \to N^-\backslash H_\beta^{\mathbb{C}}/H_\beta^{\mathbb{C}}(\Gamma)$$

be the natural projection. Then $q^!\hat{L}_\chi = \tilde{L}_\chi$ (here $q^!(\hat{L}_\chi) = \{(z,\ell) \mid z \in H_\beta^{\mathbb{C}}/H_\beta^{\mathbb{C}}(\Gamma),\ \ell \in (\hat{L}_\chi)_{q(z)}\}$). Define

$$\tilde{f}(z) = (z, f(q(z))) \quad .$$

Then $\tilde{f}$ is a holomorphic cross-section of $\tilde{L}_\chi$. Let

$$\tilde{\theta}_f(g) = g^{-1}\tilde{f}(g\cdot H_\beta^{\mathbb{C}}\Gamma)) \quad .$$

Then

$$\tilde{\theta}_f(gh) = \chi(h)^{-1}\tilde{\theta}_f(g)$$

$$\text{for } g \in H_\beta^{\mathbb{C}} \text{ and } h \in H_\beta^{\mathbb{C}}(\Gamma)$$

and

$$\tilde{\theta}_f(n^-g) = \tilde{\theta}_f(g)$$

$$\text{for } n^- \in N^-,\ g \in H_\beta^{\mathbb{C}} \quad .$$

Let

$$\theta_f(v) = \tilde{\theta}_f((P^+(v),0)) \quad .$$

If $\gamma \in \Gamma$ then

$$\begin{aligned}\theta_f((v+\gamma)) &= \tilde{\theta}_f((P^+(v+\gamma),0)) \\ &= \tilde{\theta}_f((-P^-(v+\gamma),0)(v+\gamma,\ \tfrac{1}{2}\beta(P^-(v+\gamma),\ v+\gamma))\end{aligned}$$

$$= \tilde{\theta}_f((v+\gamma,0)e^{-\pi i\beta(P^-(v+\gamma),v+\gamma)}$$

$$= \tilde{\theta}_f((v,0)(\gamma,-\tfrac{1}{2}\beta(v,\gamma))e^{-\pi i\beta(P^-(v+\gamma),u+\gamma)}$$

$$= \tilde{\theta}_f(v,0)\chi(\gamma)^{-1}e^{\pi i\beta(v,\gamma)}\ e^{-\pi i\beta(P^-(v+\gamma),v+\gamma)}$$

$$= \tilde{\theta}_f((P^-(v),0)(P^+(v),-\tfrac{1}{2}\beta(P^-(v),P^+(v))$$
$$\cdot\chi(\gamma)^{-1}e^{\pi i\beta(v,\gamma)}\ e^{-\pi i\beta(P^-(v+\gamma),v+\gamma)}$$

$$= \theta_f(v)e^{\pi i\beta(P^-(v),P^+(v))}\ \chi(\gamma)^{-1}e^{\pi i\beta(v,\gamma)}\ e^{-\pi i\beta(P^-(v+\gamma),v+\gamma)}$$

$$= \theta_f(v)\chi(\gamma)^{-1}\ \exp(2\pi i(\beta(P^-(\gamma),P^+(\gamma))/2 - \beta(P^+(v),P^-(\gamma)))\quad .$$

If we pull back $< , >$ to V we get a translation invariant Hermitian structure on V. Looking at $< , >_0$ and $T(V)_0 = V$, we find $\operatorname{Im} < , >_0 = \beta$. Let us denote $< , >_0$ by $< , >$. Then

$$\langle v,w\rangle = \beta(Jv,w) + i\beta(v,w) \quad .$$

Now

$$\beta(P^-(\gamma),P^+(\gamma)) = \frac{1}{4}\,\beta(\gamma+iJ\gamma,\ \gamma-iJ\gamma)$$
$$= \frac{1}{4}\,(-i\beta(\gamma,J\gamma) + i\beta(J\gamma,\gamma))$$
$$= \frac{i}{2}\,\langle\gamma,\gamma\rangle\ .$$

$$\begin{aligned}\beta(P^+(v),P^-(\gamma)) &= \frac{1}{4}\,\beta(v-iJv,\ \gamma+iJ\gamma)\\ &= \frac{1}{4}\,(\beta(v,\gamma) + i\beta(v,J\gamma) - i\beta(Jv,\gamma) + \beta(v,\gamma))\\ &= -\frac{i}{2}\,\langle v,\gamma\rangle \quad .\end{aligned}$$

Thus we find that if $v \in V$, $\gamma \in \Gamma$

$$\theta_f(v+\gamma) = \theta_f(v)\chi(\gamma)^{-1} \exp \pi(-\langle\gamma,\gamma\rangle/2 + \langle v,\gamma\rangle) \quad .$$

Finally $\chi(\gamma) = \exp \pi i F(\gamma)$ where F is an arbitrary function from $\Gamma \to \mathbb{R}$ satisfying

$$F(\gamma+\delta) - F(\gamma) - F(\delta) - \beta(\gamma,\delta) \in 2\mathbb{Z} \quad .$$

Hence

$$\theta_f(v+\gamma) = \theta_f(v) \exp\left(\pi \left(- \frac{\langle\gamma,\gamma\rangle}{2} + \langle v,\gamma\rangle - iF(\gamma)\right)\right)$$

a classical θ-function.

We will come back to this example in the appendix to Chapter 4 on theta functions.

Chapter 4

FOURIER ANALYSIS

In this chapter we give a rapid introduction to the Fourier analysis that we will need in the remainder of these lectures. In the first section we give a rapid introduction to the Fourier integral (here we have been influenced by the discussion in Lang [17]). In Section 2 we (briefly) study tempered distributions. We derive a formula for the Fourier transform of a general Gaussian which is quite useful to the method of stationary phase and which will be necessary to our development of the metaplectic representation. In Section 3 we study projective representations of $\mathbb{R}^{2n}$ (we call them b-representations). We give a proof of the Stone-Von Neumann theorem following the lovely exposition of Simon [23]. In Section 5 we give a classification of the irreducible unitary representations of H_n (see Chapter 3, Section 4). We also include an appendix on theta functions which completes the circle of ideas (due to Cartier [5]) in the appendix to Chapter 3.

1. THE FOURIER INTEGRAL (A RAPID REVIEW)

Let for $x \in \mathbb{R}^n$, $p = (p_1,\ldots,p_n)$, $p_i \geq 0$, $p_i \in \mathbb{Z}$, (i.e., $p \in (\mathbb{Z}^+)^n$),

$$x^p = x_1^{p_1} \cdots x_n^{p_n} .$$

Let

$$D^p = D_1^{p_1} \cdots D_n^{p_n} ,$$

$$D_i = \frac{\partial}{\partial x_i} .$$

If $p,q \in (\mathbb{Z}^+)^n$ and $f \in C^\infty(\mathbb{R}^n)$ define

$$\rho_{p,q}(f) = \sup_{x \in \mathbb{R}^n} |x^p D^q f(x)| .$$

Let $\mathcal{S}(\mathbb{R}^n)$ denote the set of all $f \in C^\infty(\mathbb{R}^n)$ such that

$$\rho_{p,q}(f) < \infty$$

for all $p,q \in (\mathbb{Z}^+)^n$.

We topologize $\mathcal{S}(\mathbb{R}^n)$ by saying that

$$\lim_{k\to\infty} f_k = f$$

if and only if

$$\lim_{k\to\infty} \rho_{p,q}(f-f_k) = 0$$

for all $p,q \in (\mathbb{Z}^+)^n$.

This topology can be given by a metric as follows: let $\xi_1,\xi_2,\ldots$ be an enumeration of $(\mathbb{Z}^+)^n \times (\mathbb{Z}^+)^n$. Define, for each ξ_i,

$$d_i(f,g) = \frac{\rho_{\xi_i}(f-g)}{1+\rho_{\xi_i}(f-g)} , \qquad i = 1,\ldots$$

Set

$$d(f,g) = \sum_{i=1}^{\infty} 2^{-i} d_i(f,g) .$$

If $f \in \mathcal{S}(\mathbb{R}^n)$ define

$$\hat{f}(x) = \frac{1}{(2\pi)^{n/2}} \int_{\mathbb{R}^n} f(y)e^{-i<x,y>} \, dy$$

($<x,y> = \Sigma x_i y_i$). We also use the notation $\hat{f} = \mathcal{F}(f)$. Now

$$|f(y)| \leq C(1+y_1^2)^{-1} \cdots (1+y_n^2)^{-1}$$

hence $\hat{f}$ is a continuous function on $\mathbb{R}^n$.

Lemma 1.1. If $f \in \mathcal{S}(\mathbb{R}^n)$ then $\hat{f} \in \mathcal{S}(\mathbb{R}^n)$ and $\mathcal{F}: \mathcal{S}(\mathbb{R}^n) \to \mathcal{S}(\mathbb{R}^n)$ is a continuous map.

Proof. We note that if $f \in \mathcal{S}(\mathbb{R}^n)$ then $\hat{f}$ is of class C^1 and

$$\frac{\partial \hat{f}}{\partial x_1}(x) = \frac{-i}{(2\pi)^{n/2}} \int_{\mathbb{R}^n} (y_i f)(y)e^{-i<x,y>} \, dy .$$

In fact,

$$\frac{\hat{f}(x+te_i) - \hat{f}(x)}{t} = \frac{1}{(2\pi)^{n/2}} \int_{\mathbb{R}^n} f(y) \frac{e^{-i<x+te_i,y>} - e^{-i<x,y>}}{t} \, dy$$

$$\left| \frac{\hat{f}(x+te_i) - \hat{f}(x)}{t} + \frac{i}{(2\pi)^{n/2}} \int_{\mathbb{R}^n} (y_i f)(y) e^{-i<x,y>} \, dy \right|$$

$$= \frac{1}{(2\pi)^{n/2}} \left| \int_{\mathbb{R}^n} f(y) e^{i<x,y>} \left(\frac{e^{-ity_i} - 1}{t} + iy_i \right) dy \right|$$

$$\leq \frac{1}{(2\pi)^{n/2}} \int_{\mathbb{R}^n} |f(y)| \left| \frac{\sin(ty_i/2)}{t/2} \quad y_i \right| dy$$

Now $|\sin x - x| \leq x^2/2$, $x \in \mathbb{R}$. Hence

$$\left| \frac{\hat{f}(x+te_i) - \hat{f}(x)}{t} + \frac{i}{(2\pi)^{n/2}} \int_{\mathbb{R}^n} (y_1 f)(y) e^{-i<x,y>} \, dy \right|$$

$$\leq |t| \frac{1}{4(2\pi)^{n/2}} \int_{\mathbb{R}^n} y_1^2 |f(y)| \, dy$$

$$\leq C|t| \quad .$$

Hence the result. But this implies that $\hat{f} \in C^{\infty}(\mathbb{R}^n)$ by repeated applications.

We have also shown:

1) If $f \varepsilon S(\mathbb{R}^n)$, $\hat{f} \varepsilon C^\infty(\mathbb{R}^n)$ and

$$D^p\hat{f}(x) = (-i)^p \frac{1}{(2\pi)^{n/2}} \int_{\mathbb{R}^n} (y^p f)(y) e^{-i<x,y>} dy$$

By a similar argument we see

2) If $f \varepsilon S(\mathbb{R}^n)$ then

$$x^p\hat{f}(x) = \frac{(i)^p}{(2\pi)^{n/2}} \int_{\mathbb{R}^n} (D^p f)(y) e^{i<x,y>} dy \quad .$$

To see this we integrate by parts.

The continuity follows from

$$\rho_{p,q}(\hat{f}) = C \sup_{\chi \varepsilon \mathbb{R}^n} \left| \int_{\mathbb{R}^n} (x^q D^p f(y) e^{-i<y,x>} dy \right|$$

$$\leq C^1 \sum_{q' \leq q+2(1,\ldots,1)} \rho_{q',p}(f)$$

(Here $q \leq r$ means $q_i \leq r_i$, $i = 1,\ldots,n$),

$$C = \frac{1}{(2\pi)^{n/2}} \quad ,$$

$$C' = C \left(\int_{-\infty}^{\infty} \frac{dx}{1+x^2} \right)^n \quad .$$

If $f,g \in \mathcal{S}(\mathbb{R}^n)$ define

$$(f*g)(x) = \frac{1}{(2\pi)^{n/2}} \int_{\mathbb{R}^n} f(x-t)g(t)\,dt$$

$$= \frac{1}{(2\pi)^{n/2}} \int_{\mathbb{R}^n} f(t)g(x-t)\,dt$$

Then by differentiating under the integral sign we see

$$D^p(f*g) = D^p f*g$$

$$= f*D^p g \quad .$$

It is also easily checked that if $f,g \in \mathcal{S}(\mathbb{R}^n)$, $f*g \in \mathcal{S}(\mathbb{R}^n)$. (Indeed, it is enough to show that $|x^p f*g| \leq C_p$ for each p.) But

$$||x|| = \langle x,x\rangle^{1/2}$$

$$\leq ||x-t|| + ||t||$$

Hence

$$||x||^k \leq \sum \binom{k}{j} ||x-t||^j \; ||t||^{k-j}$$

Thus

$$||x||^k|(f*g)(x)| \leq \sum_{j=0}^{k} \int_{\mathbb{R}^n} ||t||^j|f(t)|\cdot||x-t||^{k-j}|g(x-t)|\,dt$$

$$\leq C_k \quad .$$

Lemma 1.2. $(f*g)\hat{} = \hat{f}\hat{g}$ for $f,g \in S(\mathbb{R}^n)$.

Proof.

$$(f*g)\hat{}(x) = \frac{1}{(2\pi)^n} \int_{\mathbb{R}^n} \int_{\mathbb{R}^n} f(t)g(y-t)\,dt\, e^{-i\langle y,x\rangle}\, dy$$

$$= \frac{1}{(2\pi)^n} \int_{\mathbb{R}^n \times \mathbb{R}^n} f(t)g(y-t)e^{-i\langle y,x\rangle}\, dy\, dt$$

$$= \frac{1}{(2\pi)^n} \int_{\mathbb{R}^n \times \mathbb{R}^n} f(t)e^{-i\langle t,x\rangle} g(y-t)e^{-i\langle y-t,x\rangle}\, dy\, dt$$

$$= \frac{1}{(2\pi)^{n/2}} \int_{\mathbb{R}^n} f(t)e^{-i\langle t,x\rangle}\, dt$$

$$\cdot \frac{1}{(2\pi)^{n/2}} \int_{\mathbb{R}^n} g(y)e^{-i\langle y,x\rangle}\, dy$$

$$= \hat{f}(x)\hat{g}(x)$$

Lemma 1.3. Let

$$g(x) = e^{-x^2/2}$$

then

$$\int_{-\infty}^{\infty} g(x)\, dx = (2\pi)^{1/2} .$$

Proof. Let

$$I = \int_{-\infty}^{\infty} g(x)\, dx .$$

Then

$$I^2 = \int_{\mathbb{R}^2} e^{-(x_1^2+x_2^2)/2} dx_1\, dx_2$$

$$= 2\pi \int_0^{\infty} re^{-r^2/2}\, dr \qquad \text{(by the formula for integration in polar coordinates)}$$

$$= 2\pi \int_0^{\infty} e^{-r}\, dx$$

$$= 2\pi .$$

Q.E.D.

Lemma 1.4. $\hat{g}(x) = g(x), \quad x \in \mathbb{R}.$

Proof.

$$\frac{d}{dx}\hat{g}(x) = \frac{-i}{(2\pi)^{1/2}} \int_{-\infty}^{\infty} (yg(y))e^{-iyx}\, dy$$

$$= \frac{+i}{(2\pi)^{1/2}} \int_{-\infty}^{\infty} g'(y)e^{-iyx}\, dy = -x\hat{g}(x)$$

Now this says

$$\frac{d}{dx}\left(e^{x^2/2}\,\hat{g}(x)\right) = xe^{x^2/2}\,\hat{g}(x) - xe^{x^2/2}\,\hat{g}(x)$$

$$= 0 \quad .$$

Thus

$$\hat{g}(x) = \hat{g}(0)e^{-x^2/2} \quad ,$$

$$\hat{g}(0) = \frac{1}{\sqrt{2\pi}} \int_{-\infty}^{\infty} g(x)\,dx$$

$$= 1 \quad .$$

Q.E.D.

Theorem 1.5. If $f \in S(\mathbb{R}^n)$ then $(F(Ff))(x) = f(-x)$.

Proof. Let $g \in S(\mathbb{R}^n)$. Then

$$\frac{1}{(2\pi)^{n/2}} \int_{\mathbb{R}^n} \hat{f}(x)e^{-i<x,y>}\,g(x)\,dx$$

$$= \frac{1}{(2\pi)^n} \int_{\mathbb{R}^n\times\mathbb{R}^n} f(x)e^{-i<t,x>}\,e^{-i<x,y>}g(x)\,dx\,dt$$

$$= \frac{1}{(2\pi)^n} \int_{\mathbb{R}^n\times\mathbb{R}^n} f(t)e^{-i<y+t,x>}\,g(x)\,dx\,dt$$

$$= \frac{1}{(2\pi)^{n/2}} \int_{\mathbb{R}^n} f(t)\hat{g}(y+t)\, dt \quad .$$

Now, if $a > 0$ let $g_a(x) = h(ax)$, $h \in \mathcal{S}(\mathbb{R}^n)$. Then

$$\hat{g}_a(x) = \frac{1}{(2\pi)^{n/2}} \int_{\mathbb{R}^n} g_a(x)e^{i<y,x>}\, dy$$

$$= \frac{1}{(2\pi)^{n/2}} \int_{\mathbb{R}^n} h(ay)e^{i<y,x>}\, dy$$

$$= \frac{a^{-n}}{(2\pi)^{n/2}} \int_{\mathbb{R}^n} h(y)e^{i<y,a^{-1}x>}\, dy$$

$$= a^{-n}\, \hat{h}(a^{-1}x) \quad .$$

Thus

$$\frac{1}{(2\pi)^{n/2}} \int_{\mathbb{R}^n} f(x)e^{-i<x,y>}\, g_a(x)\, dx$$

$$= \frac{a^{-n}}{(2\pi)^{n/2}} \int_{\mathbb{R}^n} f(t)\hat{h}(a^{-1}(t+y))\, dt$$

$$= \frac{1}{(2\pi)^{n/2}} \int_{\mathbb{R}^n} f(at-y)\hat{h}(t)\, dt$$

after the change of variable $t+y/a \to t$. Now both sides of the above equation are continuous in a as $a \to 0$. Hence

$$\mathcal{F}(\hat{f})(y)h(0) = f(-y)(\mathcal{F}^2 h)(0) .$$

Take

$$h(x) = e^{-||x||^2/2}$$

then $(\mathcal{F}h)(x) = h(x)$ and $h(0) = 1$. The result now follows.

Q.E.D.

Let

$$||f||^2 = \int_{\mathbb{R}^n} |f(x)|^2 \, dx$$

for $f \in \mathcal{S}(\mathbb{R}^n)$.

<u>Theorem 1.6</u>. If $f \in \mathcal{S}(\mathbb{R}^n)$ then $||\mathcal{F}f|| = ||f||$.

<u>Proof</u>.

$$||f||^2 = \int_{\mathbb{R}^n} |f(x)|^2 \, dx$$

$$= (f_* f^*)^\wedge(0) .$$

$(f^*(x) = f(-x).)$ Then

$$\mathcal{F}(f*g) = \hat{f}\hat{g} .$$

Thus

$$F^{-1}(\hat{f}\hat{g}) = f*g \quad .$$

This says

$$F(fg)(x) = (\hat{f}*\hat{g})(x) \quad .$$

Hence

$$||f||^2 = \hat{f}*\hat{f}(0) = \int (Ff)(x)(Ff^*)(x)\, dx \quad .$$

Now

$$(Ff^*)(x) = \frac{1}{(2\pi)^{n/2}} \int_{\mathbb{R}^n} f^*(y)e^{-i<y,x>}\, dx$$

$$= \frac{1}{(2\pi)^{n/2}} \int_{\mathbb{R}^n} f(y)e^{i<y,x>}\, dx$$

$$= (Ff)(x)$$

Q.E.D.

2. TEMPERED DISTRIBUTIONS

Definition 2.1. A continuous linear map $T: S(\mathbb{R}^n) \to \mathbb{C}$ is called a tempered distribution on $\mathbb{R}^n$. Let $S'(\mathbb{R}^n)$ denote the space of all tempered distributions on $\mathbb{R}^n$.

Definition 2.2. If $f \in C^{\infty}(\mathbb{R}^n)$ we say that f is slowly increasing if there exists constants k and $C > 0$ so that $|f(x)| \leq C(1 + \|x\|)^k$.

If f is slowly increasing, we define

$$T_f(\varphi) = \int_{\mathbb{R}^n} f(x)\varphi(x)\, dx \quad ,$$

$$\varphi \in S(\mathbb{R}^n) \quad .$$

Clearly, $T_f \in S'(\mathbb{R}^n)$.

If $T \in S'(\mathbb{R}^n)$ we define $\hat{T}(\varphi) = T(\hat{\varphi})$ for $\varphi \in S(\mathbb{R}^n)$.

Lemma 2.3. If $f \in S(\mathbb{R}^n)$ then $\hat{T}_f = T_{\hat{f}}$.

Proof.

$$\hat{T}_f(\varphi) = \int_{\mathbb{R}^n} f(x)\hat{\varphi}(x)\, dx$$

$$= \frac{1}{(2\pi)^{n/2}} \int_{\mathbb{R}^n} \varphi(y) e^{-i<y,x>}\, dy\, dx$$

$$= \frac{1}{(2\pi)^{n/2}} \int_{\mathbb{R}^n \times \mathbb{R}^n} f(x)\varphi(y) e^{-i<y,x>}\, dy\, dx$$

(Fubini's theorem)

$$= \frac{1}{(2\pi)^{n/2}} \int_{\mathbb{R}^n} \varphi(y) \int_{\mathbb{R}^n} f(x) e^{-i<y,x>}\, dx\, dy$$

$$= \int_{\mathbb{R}^n} \hat{f}(x)\varphi(x)\, dx \quad .$$

Q.E.D.

Definition 2.4. If $f \in C^\infty(\mathbb{R}^n)$ is slowly increasing and if there exists h a slowly increasing function on $\mathbb{R}^n$ so that $\hat{T}_f = T_h$, then we use the notation $\hat{f} = h$.

Note. $\hat{f}$ is well defined if it exists. Lemma 2.3 says that our notation is consistent.

Lemma 2.5. Let A be a symmetric nonsingular $n \times n$ matrix. Define

$$g_A(x) = e^{i\langle Ax,x\rangle/2} \quad .$$

Then

$$\hat{g}_A(x) = |\det A|^{-1/2}\, e^{(\pi i/4)\operatorname{sgn} A}\, e^{-i\langle A^{-1}x,x\rangle/2} \quad .$$

Here $\operatorname{sgn}(A)$ is the number of positive eigenvalues of A minus the number of negative eigenvalues of A.

Proof. Let $v_1,\ldots,v_n$ be an orthonormal basis of $\mathbb{R}^n$ so that $Av_i = \lambda_i v_i$, $i = 1,\ldots,n$. Then $\lambda_i \neq 0$ for $i = 1,\ldots,n$. Now, using $v_1,\ldots,v_n$ as "standard basis" we have

$$\hat{T}_{g_A}(f) = \int_{\mathbb{R}^n} e^{i\,\Sigma\lambda_i x_i^2/2}\, \hat{f}(x)\, dx \quad .$$

Thus if we show for $n = 1$ that if $\lambda \neq 0$, and $g_\lambda(x) = e^{i\lambda x^2/2}$ that

$$\hat{g}_\lambda(x) = |\lambda|^{-1/2} e^{\pi i/4 \, \mathrm{sgn}(\lambda)} e^{-i\lambda^{-1}x^2/2}$$

we will be done. Set

$$h_z(x) = e^{-zx^2/2}$$

for $z \in \mathbb{C}$.

If $z \in \mathbb{C}$, $\mathrm{Re}\, z > 0$, then $h_z \in \mathcal{S}(\mathbb{R})$.

1) If $x \in \mathbb{R}$ then $z \to \hat{h}_z(x)$ is holomorphic in z for $\mathrm{Re}\, z > 0$.

This is seen by noting that

$$\frac{1}{(2\pi)^{1/2}} \int_{-\infty}^{\infty} e^{-zx^2/2} e^{-iyx} \, dx$$

converges uniformly for z in a compact subset of $\mathrm{Re}\, z > 0$.

2) If $\mathrm{Re}\, z > 0$ then

$$\hat{h}_z(x) = \frac{1}{\sqrt{z}} e^{-x^2/2z}$$

Here we use the branch of $\sqrt{z}$ on $\mathrm{Re}\, z > 0$ that agrees with the positive square root on $x > 0$, $x \in \mathbb{R}$.

To prove 2) we need only note that 2) is true for $z \in \mathbb{R}$, $z > 0$. But, $h_z(x) = h_1(\sqrt{z}x)$. $\hat{h}_1(x) = h_1(x)$. Hence

$$\hat{h}_z(x) = \frac{1}{\sqrt{z}} h_1(x/\sqrt{z})$$

$$= \frac{1}{\sqrt{z}} e^{-x^2/2z} .$$

This proves 2).

We extend $z \to z^{-1/2}$ to $\operatorname{Re} z \geq 0$, $z \neq 0$ as follows:

$$z = Re^{i\theta} , \quad -\frac{\pi}{2} \leq \theta \leq \frac{\pi}{2} ,$$

$$z^{-1/2} = R^{-1/2} e^{-i\theta/2} .$$

Now, if $z \in \mathbb{R}$,

$$\lim_{\substack{x\to 0\\ x>0}} \hat{h}_{-iz+x}(\xi) = \lim_{\substack{x\to 0\\ x>0}} (x-iz)^{-1/2} e^{-\xi^2/2(x-iz)}$$

$$= |z|^{-1/2} e^{\pi i/4 \operatorname{sgn} z} e^{+\xi^2/2iz}$$

$$= |z|^{-1/2} e^{\pi i/4 \operatorname{sgn} z} e^{-i\xi^2/2z}$$

But by using the Lebesgue dominated convergence theorem, we have

$$\hat{T}_{g_z}(f) = \lim_{\substack{x\to 0\\ x>0}} \int_{-\infty}^{\infty} e^{(iz-x)t^2/2} \hat{f}(t)\, dt$$

$$= \lim_{\substack{x\to 0\\ x>0}} \int_{-\infty}^{\infty} \hat{h}_{-iz+x}(t) f(t)\, dt$$

$$= \int_{-\infty}^{\infty} |z|^{-1/2} e^{\pi i/4(\mathrm{sgn}\ z)} e^{-it^2/2z} f(t)\, dt .$$

Q.E.D.

3. THE STONE-VON NEUMANN THEOREM

If $f \in \mathcal{S}(\mathbb{R}^n)$, $v \in \mathbb{R}^n$ define

$$S(v)f(x) = e^{i\langle v,x\rangle} f(x) .$$

Then

$$||S(v)f|| = ||f||$$

for $f \in \mathcal{S}(\mathbb{R}^n)$.

Hence $S(v)\colon L^2(\mathbb{R}^n) \to L^2(\mathbb{R}^n)$ is a bounded linear operator.

Lemma 3.1. Suppose that $A\colon L^2(\mathbb{R}^n) \to L^2(\mathbb{R}^n)$ is a bounded linear operator such that

$$A \circ S(v) = S(v) \circ A$$

for all $v \in \mathbb{R}^n$.

Set

$$g(x) = e^{-||x||^2/2} .$$

Then if $f \in \mathcal{S}(\mathbb{R}^n)$,

$$Af = g^{-1}(Ag) \cdot f .$$

Proof. If $\alpha \in \mathcal{S}(\mathbb{R}^n)$ then the Riemann sums converging to $\hat{\alpha}(x)$ converge uniformly for $x \in \mathbb{R}^n$. This implies that

$$\frac{1}{(2\pi)^{n/2}} \int_{\mathbb{R}^n} \alpha(v)S(v)\, dv \cdot f = \hat{\alpha}f ,$$

for $f \in \mathcal{S}(\mathbb{R}^n)$.

The continuity of A now implies that

$$A(\hat{\alpha}f) = \hat{\alpha}Af$$

for $f \in \mathcal{S}(\mathbb{R}^n)$.

But then

$$A(\alpha f) = \alpha A(f)$$

for $\alpha, f \in \mathcal{S}(\mathbb{R}^n)$.

In particular, $A(gf) = (Ag)f$ and $A(gf) = gAf$. Hence

$$Af = (g^{-1}(Ag))\cdot f .$$

Q.E.D.

Define for $f \in \mathcal{S}(\mathbb{R}^n)$, $v \in \mathbb{R}^n$, $(T(v)f)(x) = f(x-v)$. Clearly $\|T(v)f\| = \|f\|$. Hence $T(v)$ extends to a bounded operator on $L^2(\mathbb{R}^n)$.

Corollary 3.2. If $A: L^2(\mathbb{R}^n) \to L^2(\mathbb{R}^n)$ is a bounded linear operator and if $A \circ S(v) = S(v) \circ A$, $A \circ T(v) = T(v) \circ A$, for all $v \in \mathbb{R}^n$, then $A = cI$, $c \in \mathbb{C}$.

Proof. Set

$$h = g^{-1}(Ag) = g^{-1}u ,$$

$$u \in L^2(\mathbb{R}^n) .$$

(See Lemma 3.1.) Now

$$T(v)Af = (T(v)h)(T(v)f) .$$

Taking $f = g$ we see $T(v)h = h$, h a measurable function on $\mathbb{R}^n$. But then h is equal to a constant, c, almost everywhere. Hence $Af = cf$, for $f \in \mathcal{S}(\mathbb{R}^n)$. Q.E.D.

We look at $\mathbb{R}^{2n}$ as the space of all columns

$$\begin{bmatrix} x \\ y \end{bmatrix}, \quad x,y \in \mathbb{R}^n .$$

We will also think of $\mathbb{R}^n$ as $n \times 1$ columns. Set

$$b\left(\begin{bmatrix} x \\ y \end{bmatrix}, \begin{bmatrix} x' \\ y' \end{bmatrix}\right) = e^{-i\langle x',y\rangle} .$$

We define

$$U\left(\begin{bmatrix} x \\ y \end{bmatrix}\right)f = S(x)T(y)f$$

$$\text{for } f \in L^2(\mathbb{R}^n) .$$

A direct computation gives

Lemma 3.3. $U(x)U(w) = b(z,w)U(z+w)$.

By a <u>Hilbert space</u> we will mean a separable, complete inner product space over $\mathbb{C}$. If $(H, \langle\ ,\ \rangle)$ is a Hilbert space we denote by $U(H)$ the group of all unitary bijections of H to H. We will use the <u>strong topology</u> on $U(H)$. That is, $U_j \to U_0$ if and only if $U_j\varphi \to U_0\varphi$ for all $\varphi \in H$.

Let $b\colon \mathbb{R}^{2n} \times \mathbb{R}^{2n} \to \mathbb{C}$ be as above.

<u>Definition 3.4</u>. A b-representation of $\mathbb{R}^{2n}$ on $(H, \langle\ \rangle)$ is a strongly continuous map $u\colon \mathbb{R}^{2n} \to U(H)$ so that

$$u(x)u(y) = b(x,y)u(x+y) \quad .$$

<u>Example 3.5</u>. $(U, L^2(\mathbb{R}^n))$.

<u>Definition 3.6</u>. A b-representation of $\mathbb{R}^{2n}$, (u,H) is said to be irreducible if the only bounded linear operator A mapping H to H so that $A \circ u(x) = u(x) \circ A$ for all $x \in \mathbb{R}^{2n}$ are scalar multiples of the identity.

<u>Example 3.7</u>. Corollary 3.2 says that $(U, L^2(\mathbb{R}^n))$ is irreducible.

Recall that if $(H_i, \langle\ ,\ \rangle_i)$, $i = 1,2$ are Hilbert spaces then $H_1 \hat{\otimes} H_2$ is the Hilbert space completion of $H_1 \otimes H_2$ (algebraic tensor product) relative to $\langle v_1 \otimes v_2, w_1 \otimes w_2\rangle = \langle v_1, w_1\rangle_1 \langle v_2, w_2\rangle_2$.

<u>Definition 3.8</u>. Let (u_i,H_i), $i = 1,2$ be b-representations. Then (u_1,H_1) is said to be <u>equivalent</u> with (u_2,H_2) if there is a bijective unitary operator $A: H_1 \to H_2$ so that $A\circ u_1(x) = u_2(x)\circ A$ for all $x \in \mathbb{R}^{2n}$.

(u_1,H_1) is said to be <u>quasi-equivalent</u> with (u_2,H_2) if there exist Hilbert spaces H_1' and H_2' so that if

$$u_i'(x)(\psi \otimes z) = u_i(x)\psi \otimes z ,$$

$$x \in \mathbb{R}^{2n}, \quad \psi \in H_i, \quad z \in H_i'$$

then $(u_1', H_1 \hat{\otimes} H_1')$ is equivalent with $(u_2', H_2 \hat{\otimes} H_2')$.

<u>Lemma 3.9</u> (Fell). Let

$$(B(x)f)(y) = b(y,x)f(y+x)$$

$$\text{for } f \in L^2(\mathbb{R}^{2n}), \quad x \in \mathbb{R}^{2n} .$$

Then B is a b-representation. Let (u,H) be a b-representation of $\mathbb{R}^{2n}$. Let $(R, L^2(\mathbb{R}^{2n}))$ be the representation $(R(x)f) = f(y+x)$. Then $u \otimes R$ is quasi-equivalent with B.

<u>Proof</u>. Let us identify $H \hat{\otimes} L^2(\mathbb{R}^{2n})$ with $L^2(\mathbb{R}^{2n}; H)$ the Hilbert space of all $f: \mathbb{R}^{2n} \to H$ such that

$$\int_{\mathbb{R}^{2n}} \langle f(x),f(x)\rangle \, dx < \infty$$

under the map $(v \otimes f)(x) = f(x)v$. Let

$$A\colon L^2(\mathbb{R}^{2n};H) \to L^2(\mathbb{R}^{2n};H)$$

be defined by

$$(Af)(x) = u(x)f(x) \quad .$$

Then $||Af|| = ||f||$. Thus A is a unitary operator. Clearly,

$$(A^{-1}f)(x) = \frac{u(x)^{-1}}{b(x,x)} f(x) \quad .$$

Now,

$$((u\otimes R)(x)f)(y) = u(x)\cdot f(x+y)$$

$$\text{for} \quad f \in L^2(\mathbb{R}^{2n};H)$$

relative to the above identification.

Now

$$\begin{aligned}(A\circ(u\otimes R)(x)f)(y) &= U(y)((u\otimes R)(x)f)(y)\\ &= u(y)u(x)f(x+y)\\ &= b(y,x)u(x+y)f(x+y)\\ &= b(y,x)(Af)(x+y)\\ &= (B(x)\otimes I)(Af)(y) \quad .\end{aligned}$$

Thus $u \otimes R$ is equivalent with $B \hat{\otimes} I_H$. Q.E.D.

<u>Lemma 3.10</u>. If (U_i,H_i), $i = 1,2$ are irreducible, quasi-equivalent b-representations of $\mathbb{R}^{2n}$ then (U_1,H_1) and (U_2,H_2) are equivalent.

<u>Proof</u>. Let

$$A\colon H_1 \hat{\otimes} H_3 \to H_2 \hat{\otimes} H_4$$

be such that

$$A\circ(U_1(x)\otimes I) = (U_2(x)\otimes I)\circ A$$
$$x \in \mathbb{R}^{2n} ,$$

A a bijective unitary operator. Let $v_1, v_2, \ldots$ be an orthonormal basis of H_4. Let

$$P_i\colon H_2 \otimes H_4 \to H_2 \otimes H_4$$

be defined by

$$P_i(v\otimes w) = \langle w, v_i\rangle\, v \otimes v_i \quad .$$

Then

$$\|P_i(z)\| \leq \|z\|$$
$$\text{for} \quad z \in H_1 \otimes H_4 \quad .$$

Hence

$$P_i\colon H_1 \hat{\otimes} H_4 \to H_1 \hat{\otimes} H_4$$

is a bounded operator.

a) If $P_i(z) = 0$ for $i = 1,2,\ldots$ then $z = 0$.

This is obvious since $H_1 \hat{\otimes} H_4$ has an orthonormal basis of the form $\phi_j \otimes v_i$.

Let $Q_i: H_2 \otimes H_4 \to H_2$ be defined by

$$Q_i(v \otimes w) = \langle w, v_i \rangle v \quad .$$

Then $\|Q_i(z)\| \leq \|z\|$, as above. Hence $Q_i: H_2 \hat{\times} H_4 \to H_2$ is a bounded operator. Furthermore, $Q_i: P_i(H_2 \hat{\otimes} H_2) \to H_2$ is a unitary operator.

Let $v_0 \in H_3$ be a unit vector. Let i be so that

$$P_i(A(\psi \otimes v_0)) \not\equiv 0$$

for $\psi \in H_1$.

Then set

$$S(\psi) = Q_i P_i A(\psi \otimes v_0) \quad .$$

Then $S: H_1 \to H_2$, $S \not\equiv 0$. Furthermore,

$$\begin{aligned} S(U_1(x)\psi) &= Q_i P_i A(U_2(x)\psi \otimes v_0)) \\ &= Q_i P_i (U_2(x) \otimes I) A(\psi \otimes v_0) \\ &= U_2(x) S(\psi) \quad . \end{aligned}$$

b) S is injective

Indeed, if

$$V = \{\psi \in H_1 \mid S(\psi) = 0\}$$

then $U(x)V \subset V$ for $x \in \mathbb{R}^{2n}$. Since $U(x)$ is unitary,

$U(x)V^{\perp} \subset V^{\perp}$ $(V^{\perp} = \{\psi \in H_1 | \langle\psi,V\rangle = 0\}$. Thus if $P: H_1 \to V$ is the projective relative to $H_1 = V \oplus V^{\perp}$,

$$P \circ U_1(g) = U_1(g) \circ P$$

for $g \in \mathbb{R}^{2n}$.

Hence $P = \lambda I$. But if $\lambda = 0$, $V = 0$ if $\lambda \neq 0$, $V = H_1$. Since $V \neq H_1$ we see $V = 0$.

c) S is surjective

This time $S(H_1) = V$ and as above $P: H_2 \to V$. $P \circ U_2(g) = U_2(g) \circ P$. $P = \lambda I$. $\lambda \neq 0$ since $S(H_1) \neq 0$, hence $V = H_2$.

Now defining

$$\langle S^*\psi,\phi\rangle = \langle\psi,S(\phi)\rangle$$

for $\psi,\phi \in H_2$

we see $S^*S: H_1 \to H_1$ is bounded and

$$S^*S \circ U_1(x) = U_1(x) \circ S^*S$$

for $x \in \mathbb{R}^{2n}$.

Hence $S^*S = \lambda I$. $\lambda \geq 0$. Let $W = (1/\sqrt{\lambda})S$. Then W is the desired unitary bijection

$$W \circ U_1(g) = U_2(g) \circ W \quad ,$$

$g \in \mathbb{R}^{2n}$.

Q.E.D.

Theorem 3.11 (Stone-Von Neumann). If (u,H) is an irreducible b-representation of $\mathbb{R}^{2n}$ then (u,H) is equivalent with $(U, L^2(\mathbb{R}^n))$.

Proof. It is enough to prove that if (u,H) is a b-representation of $\mathbb{R}^{2n}$ then (u,H) is quasi-equivalent with $u \otimes R$. Indeed, if we show this, then (u,H) and $(U, L^2(\mathbb{R}^n))$ are quasi-equivalent with $(B, L^2(\mathbb{R}^{2n}))$, (Lemma 3.9), hence with each other. The theorem then follows from Lemma 3.10. We look upon $H \hat{\otimes} L^2(\mathbb{R}^{2n})$ as $L^2(\mathbb{R}^{2n}; H)$ as above. Let $J: \mathbb{R}^{2n} \to \mathbb{R}^{2n}$ be defined by

$$J \begin{bmatrix} x \\ y \end{bmatrix} = \begin{bmatrix} y \\ -x \end{bmatrix} .$$

Let $\mathcal{F}: L^2(\mathbb{R}^{2n}) \to L^2(\mathbb{R}^{2n})$ be the Fourier transform. Then

$$\mathcal{F} \circ R(x) \circ \mathcal{F}^{-1} f = \chi_x f$$

where

$$\chi_x(y) = e^{i\langle x,y\rangle} , \qquad \langle x,y\rangle = \sum_{i=1}^{n} x_i y_i .$$

Hence $(R, L^2(\mathbb{R}^{2n}))$ is equivalent with $(\hat{R}, L^2(\mathbb{R}^{2n}))$, $\hat{R}(x)f = \chi_x f$. Thus we must show $u \otimes \hat{R}$ is quasi equivalent with u. We look at

$$((u \otimes \hat{R})(x)f)(z) = \chi_x(z) u(x) \circ f(z)$$

$$f \in L^2(\mathbb{R}^{2n}; H) .$$

Define $(Cf)(z) = u(z)^{-1} \cdot f(Jz)$.

1) C defines a bijective unitary operator.

This is clear, since $|\det J| = 1$.

2) If $x,y \in \mathbb{R}^{2n}$ then $u(y)^{-1}u(x) = \chi_{Jy}(x)u(x)u(y)^{-1}$.

To see this let

$$x = \begin{bmatrix} x_1 \\ x_2 \end{bmatrix}, \qquad y = \begin{bmatrix} y_1 \\ y_2 \end{bmatrix} .$$

$$u(y)^{-1} = e^{i<y_2,y_1>} u(-y) .$$

Hence

$$u(y)^{-1}u(x) = e^{i<y_2,y_1>} b(-y,x)u(x-y)$$

$$u(x)u(y)^{-1} = e^{i<y_2,y_1>} b(x,-y)u(x-y)$$

Thus

$$u(y)^{-1}u(x) = \frac{b(-y_1,x)}{b(x,-y)} u(x)u(y^{-1})$$

$$\frac{b(-y,x)}{b(x,-y)} = \frac{e^{i<x_1,y_2>}}{e^{i<y_1,x_2>}} = e^{i(<x_1,y_2> - <x_2,y_1>)}$$

$$= e^{i<x,Jy>} = \chi_{Jy}(x) .$$

This prove (2).

Now

$$
\begin{aligned}
((C\circ(\hat{R}\otimes u(x))f)(y) &= U(y)^{-1}((\hat{R}\otimes u)(x)f)(Jy) \\
&= u(y)^{-1}\chi_x(Jy)u(x)f(Jy) \\
&= \chi_{Jy}(x)u(y)^{-1}(x)f(Jy) \\
&= u(x)u(y)^{-1}f(Jy) \\
&= (I\otimes u(x))(Cf)(y) \quad .
\end{aligned}
$$

Hence u is quasi-equivalent with $\hat{R}\otimes u$. Q.E.D.

We actually proved a bit more than the above theorem:

Corollary 3.13 (to the Proof of Theorem 3.12). If U_1, U_2 are b-representations, then they are quasi-equivalent.

Theorem 3.14. Let $(H, < >)$ be a Hilbert space. Let $(U_H, L^2(\mathbb{R}^n; H))$ be the b-representation of $\mathbb{R}^{2n}$ given by

$$\left(U_H\left(\begin{bmatrix} x \\ y \end{bmatrix}\right)f\right)(z) = e^{i<z,x>} f(z-y) \quad .$$

If (u,V) is a b-representation of $\mathbb{R}^{2n}$ then (u,V) is equivalent with $(U_H, L^2(\mathbb{R}^n; H))$ for H some Hilbert space.

Note. $U_{\mathbb{C}} = U$.

The proof of this result depends on the following simple lemma.

Lemma 3.15. Let $\xi: H^* \times H \to \mathbb{C}$ (H^* the continuous dual space of H. That is, $H^* = H$ and $\lambda \cdot \varphi = \bar{\lambda}\varphi$) be bilinear and $|\xi(\lambda,\varphi)| \leq C\|\lambda\| \; \|\varphi\|$, for C fixed and $\lambda \in H^*$, $\varphi \in H$ arbitrary. Then there exists $A: H \to H$ linear and bounded ($\|Av\| \leq C\|v\|$) so that $\xi(\lambda,\varphi) = \lambda(A\varphi)$.

Proof. If $\lambda \in H^*$ is fixed then the map $\varphi \mapsto \xi(\lambda,\varphi)$ is continuous. Hence

$$\xi(\lambda,\varphi) = \langle\varphi, Q(\lambda)\rangle ,$$

$$Q(\lambda) \in H .$$

Now

$$|\xi(\lambda, Q(\lambda))| \leq \|\lambda\| \; \|Q(\lambda)\|C .$$

Hence

$$\|Q(\lambda)\|^2 \leq C\|\lambda\| \; \|Q(\lambda)\|$$

This implies that $\|Q(\lambda)\| \leq C\|\lambda\|$. Hence $Q: H^* \to H$ is bounded, anti-linear, and $\xi(\lambda,\varphi) = \langle\varphi, Q(\lambda)\rangle$. If $\varphi_\lambda \in H$ is defined by $\lambda(\varphi) = \langle\varphi,\varphi_\lambda\rangle$, then the map $H^* \to H$, $\lambda \mapsto \varphi_\lambda$ is continuous and anti-linear. Let

$$\tilde{Q}(\varphi_\lambda) = Q(\lambda) .$$

Then $\tilde{Q}: H \to H$ is linear and bounded. Set $A = \tilde{Q}^*$. We then have

$$\xi(\lambda,\varphi) = \langle A\varphi,\varphi_\lambda\rangle = \lambda(A\varphi) .$$

Q.E.D.

Proof. of Theorem 3.14. We first show that if $A: L^2(\mathbb{R}^n,H) \to L^2(\mathbb{R}^n,H)$ and $A \circ U_H(x) = U_H(x) \circ A$ for all $x \in \mathbb{R}^{2n}$ then $(Af)(x) = B \cdot f(x)$, $B: H \to H$ bounded. Let for $f \in L^2(\mathbb{R}^n)$, $v \in H$, $(f \otimes v)(x) = f(x)v$. Then

$$f \otimes v \in L^2(\mathbb{R}^n,H) \quad .$$

Now if $\lambda \in H^*$ and if $C(\lambda,v)f = (I \otimes \lambda)A(F \otimes v)$, then

$$C(\lambda,v): L^2(\mathbb{R}^n) \to L^2(\mathbb{R}^n)$$

and

$$C(\lambda,v) \circ U(x) = U(x) \circ C(\lambda,v) \quad .$$

Hence

$$C(\lambda,v) = \xi(\lambda,v)I \quad .$$

Clearly

$$|\xi(\lambda,v)| \leq ||A|| \; ||v|| \; ||\lambda|| \quad .$$

Hence

$$(I \otimes \lambda)(A(f \otimes v)) = \lambda(Bv)f \quad ,$$

$B: H \to H$ bounded. But then if $\mu \in L^2(\mathbb{R}^n)^*$ then

$$(\mu \otimes \lambda)(A(f \otimes v)) = \mu(f)\lambda(Bv) \quad .$$

This implies $(\mu \otimes \lambda) \circ (A - I \otimes B) = 0$. Hence $A = I \otimes B$, which proves the above assertion.

Let H_1,H_2 be Hilbert spaces so that $u \otimes I$ on $V \hat{\otimes} H_1$ is equivalent with U_{H_ℓ} (this is possible by Corollary 3.13).

Let $v \in H_1$ be a unit vector. Let $T: V \hat{\otimes} H_1 \to L^2(\mathbb{R}^n, H_2)$ be the equivalence. Set

$$T_v(\varphi) = T(\varphi \otimes v) \quad .$$

$T: V \to L^2(\mathbb{R}^n, H_2)$ is unitary and intertwines the representation u and U_H. If $T_v(\varphi) = 0$ then $T(\varphi \otimes v) = 0$. Hence $\varphi \otimes v = 0$. But then $\varphi = 0$. Also $|T_v(\varphi)| = ||\varphi||$. Thus $T_v(V)$ is closed in $L^2(\mathbb{R}^n, H_2)$ and U_{H_2} invariant. Let $P: L^2(\mathbb{R}^n, H_2) \to T_v(V)$ be the orthogonal projection. The above observation implies that $P = I \otimes Q$, $Q: H \to H$ a projection. Hence

$$T_v(V) = L^2(\mathbb{R}^n) \otimes Q(H_2)$$

$$= L^2(\mathbb{R}^n; Q(H_2)) \quad .$$

Thus

$$T_v: V \to L^2(\mathbb{R}^n; Q(H_2))$$

is the desired unitary equivalence. Q.E.D.

4. THE IRREDUCIBLE UNITARY REPRESENTATIONS OF THE HEISENBERG GROUP

If H is a Hilbert space we will denote by U(H) the group of all bijective unitary operators from H to H. We will always use the "strong topology" on U(H). That is, $u_j \to u_0$, $u_j \in U(H)$ if and only if $u_j \cdot \varphi \mapsto u_0 \cdot \varphi$ for all $\varphi \in H$.

If G is a topological group, then a <u>unitary representation</u> (π, H) of G is a continuous homomorphism, $\pi: G \to U(H)$. If (π_i, H_i) are unitary representations of G, $i = 1,2$, then π_1 is said to be <u>equivalent</u> with π_2 if there exists $A: H_1 \to H_2$ a unitary bijection so that

$$A \circ \pi_1(g) = \pi_2(g) \circ A$$

for all $g \in G$.

(π, H) a unitary representation of G is said to be <u>irreducible</u> if the only bounded operators $A: H \to H$ such that

$$A \circ \pi(g) = \pi(g) \circ A$$

for all $g \in G$

are the operators λI, $\lambda \in \mathbb{C}$.

In this section we give a realization up to equivalence of the set of irreducible unitary representations of H_n (see Section 3.4). Recall that H_n is the $2n+1$ dimensional Heisenberg group. We use the realization

$$H_n = \mathbb{R}^{2n} \times \mathbb{R}$$

and

$$(x,t) \cdot (y,s) = (x+y, t+s+\tfrac{1}{2}\beta(x,y)) ,$$

$$\beta\left(\begin{bmatrix} x \\ y \end{bmatrix}, \begin{bmatrix} u \\ v \end{bmatrix}\right) = (x,v) - (y,u)$$

(see Section 3.4).

Let now (π,H) be an irreducible unitary representation of H_n. Let

$$z = \{(0,t) \mid t \in \mathbb{R}\} \subset H_n .$$

Then z is the center of H_n and hence

$$\pi(0,t)\pi(g) = \pi(g)\pi(0,t)$$

$$\text{for all } t \in \mathbb{R}, \quad g \in H_n .$$

Hence $\pi(0,t) = \chi(t)I$, $\chi\colon \mathbb{R} \to T^1 = \{z \in \mathbb{C} \mid |z| = 1\}$ a continuous homomorphism. Hence

$$\chi(t) = e^{i\lambda t} ,$$

$$\lambda \in \mathbb{R} .$$

a) $\lambda = 0$. Then $\pi(v,t) = \pi(v,0)$ for all $v \in \mathbb{R}^{2n}$, $t \in \mathbb{R}$.

$$\pi(v,t)\cdot\pi(w,s) = \pi(v+w,\ t+s+\tfrac{1}{2}\beta(v,w))$$

$$= \pi(v+w,\ 0) .$$

Hence

$$\pi(g)\pi(x) = \pi(x)\pi(g)$$

$$\text{for all } x,g \in G .$$

This implies, by irreducibility, that

$$\pi(v,t) = \eta(v)I ,$$

$\eta: \mathbb{R}^{2n} \to T^1$,

$$\eta(x+y) = \eta(x)\eta(y) \quad ,$$

η continuous. Therefore, there exists $\mu:(\mathbb{R}^{2n})^*$ so that $\eta(x) = e^{i\mu(x)}$, $x \in \mathbb{R}^{2n}$ and $\dim H = 1$. We have therefore completely described the case $\lambda = 0$.

b) $\lambda \neq 0$. Let $w(v) = \pi(v,0)$. Then

$$\begin{aligned} w(v)w(v') &= \pi((v,0)(v',0)) \\ &= \pi(v+v', \tfrac{1}{2}\beta(v,v')) \\ &= e^{i\lambda/2\ \beta(v,v')}\ w(v+v') \quad . \end{aligned}$$

Let $e_1,\ldots,e_{2n}$ be a basis for $\mathbb{R}^{2n}$ satisfying

$$\begin{aligned} (\lambda\beta)(e_i,e_j) &= (\lambda\beta)(e_{n+i}, e_{n+j}) \\ &= 0 \quad , \qquad i.j \leq n \end{aligned}$$

and

$$(\lambda\beta)(e_i, e_{n+j}) = \delta_{ij} \quad .$$

Set, for $x,y \in \mathbb{R}^n$,

$$w\begin{bmatrix} x \\ y \end{bmatrix} = w(\Sigma x_i e_i + \Sigma y_i e_{n+i}) \quad .$$

Set

$$u\left(\begin{bmatrix} x \\ y \end{bmatrix}\right) = e^{i/2 \langle x,y\rangle} w\begin{bmatrix} x \\ y \end{bmatrix} .$$

It is easily checked that (u,H) is a b-representation (see Section 3). Since π is irreducible it is clear that (u,H) is irreducible. Theorem 3.11 implies that there is a bijective unitary operator $A: H \to L^2(\mathbb{R}^n)$ so that

$$A \circ u\begin{bmatrix} x \\ y \end{bmatrix} = U\left(\begin{bmatrix} x \\ y \end{bmatrix}\right) \circ A .$$

This says that

$$\left(A \circ \pi \left(\sum_{i=1}^{n} x_i e_i + \sum_{j=1}^{n} y_j e_{n+j}, t \right) \circ A^{-1} f \right)(z)$$

I)

$$= e^{i\lambda t} e^{i\lambda \langle x,\ z - \frac{1}{2}y\rangle} f(z-y) .$$

We have proved

<u>Theorem 4.1</u>. If (π,H) is an irreducible unitary representation of H_n, then $\pi(0,t) = e^{i\lambda t} I$, $\lambda \in \mathbb{R}$.

i) If $\lambda = 0$ then $\dim H = 1$ and $\pi(v,t) = e^{i\mu(v)}$, $\mu \in (\mathbb{R}^{2n})^*$.

ii) If $\lambda \neq 0$ and if we define

$$\left(S_\lambda\left(\begin{bmatrix} x \\ y \end{bmatrix}, t\right) f\right)(x) = e^{i\lambda(t + \langle x, z - \frac{1}{2}y\rangle)} f(z-y)$$

Then $(S_\lambda, L^2(\mathbb{R}^n))$ defines an irreducible unitary representation equivalent with π.

APPENDIX TO CHAPTER 4

THETA FUNCTIONS

In this section we show how to combine the results of the Appendix to Chapter 3 with the results of Chapter 4 to derive results about theta functions. The following discussion takes its ideas from Cartier's article [5].

Let V be a $2n$-dimensional vector space. Let $\Gamma \subset V$ be a discrete (additive) subgroup so that $V/\Gamma = T$ is compact.

Let $\beta\colon V \times V \to \mathbb{R}$ be a non-degenerate alternative bilinear form. In light of the results of A-3 we assume $\beta(\Gamma,\Gamma) \subset \mathbb{Z}$. Then β induces an integral symplectic structure on T.

Let us use the notation of A-3. In particular,

$$H_\beta(\Gamma) = \{(\gamma,t) \mid \gamma \in \Gamma,\ t \in \mathbb{R}\} \subset H_\beta$$

Let $\chi\colon H_\beta(\Gamma) \to T^1$ be a character of $H_\beta(\Gamma)$, so that

1) $\chi(0,t) = e^{2\pi it}, \qquad t \in \mathbb{R}$.

Then we have seen that if $F\colon \Gamma \to \mathbb{R}$ is such that

2) $F(\gamma) + F(\delta) + \beta(\gamma,\delta) - F(\gamma+\delta) \in 2\mathbb{Z}$

then

3) $\chi(\gamma,t) = e^{\pi iF(\gamma)}\, e^{2\pi it}$

defines such a character. Once we have fixed such a character, we can form a line bundle over T,

$$L_\chi \to T \quad .$$

L_χ is the line bundle gotten by looking at $H_\beta \times \mathbb{C}/H_\beta(\Gamma)$ relative to the action $(h,z)\cdot(\xi) = (h\xi, \chi(\xi)^{-1}z)$.

We have seen how to form a Hermitian structure, h, and a connection on L_χ so that

$$4) \quad 2\pi i\omega \;=\; \operatorname{curv}(L_\chi,\nabla) \quad .$$

Furthermore, H_β acts on L_χ by bundle maps and if $g \in H_\beta$, $\ell \in L_\chi$, then $h(g\cdot\ell, g\cdot\ell) = h(\ell,\ell)$.

We form a unitary representation of H_β as follows: Let

$$\langle f,g\rangle \;=\; \int_T h\langle f(x),g(x)\rangle \, dx$$

$$\text{for} \quad g,f \in C^\infty(T;\, L_\chi) \quad .$$

(Here dx can be taken to be the invariant measure on T with total mass 1.) Let

$$(T_\chi(y)f)(z) \;=\; y\cdot f(y^{-1}\cdot z)$$

$$\text{for} \quad y \in H_\beta, \quad z \in T, \quad f \in C^\infty(T;\, L_\chi) \quad .$$

Let H^χ be the Hilbert space completion of $(C^\infty(T;L_\chi), \langle\ ,\ \rangle)$. Then

$$T_\chi(y) \in U(H^\chi)$$

for $y \in H_\beta$.

Furthermore,

5) (T_χ, H^χ) defines a unitary representation of H_β.

5) is a direct consequence of, say, the results in Wallach [26] (see in particular, Theorem 2.4.6).

We note that a direct computation yields

6) $T_\chi(0,t)f = e^{2\pi it} f$, for $t \in \mathbb{R}$.

Using (8.16.8) and (2.7.4) of Wallach [26] we have

Lemma A.4.1.

$$H^\chi = \sum_{j=1}^{N} H_j$$

with H_j a closed invariant subspace with T_χ acting on H_j irreducibly (N might be ∞). Furthermore, fixing j, there are only a finite number of H_i so that H_j is equivalent with H_i.

Let

$$\pi_j(g) = T_\chi(g)|_{H_j}$$

Then (π_j, H_j) is an irreducible unitary representation of H_β. Furthermore,

$$\pi_j(0,t) = e^{2\pi it}$$

for $t \in \mathbb{R}$.

Theorem 5.1 implies that (π_j, H_j) is equivalent with the standard representation of H_β corresponding to 2π. Let us denote this representation by $S_{2\pi}$. In particular, (π_j, H_j) is equivalent with (π_i, H_i) for all i,j.

We therefore see (using Lemma A.4.1),

Lemma A.4.2. $H^\chi = H_1 \oplus \cdots \oplus H_{N_\chi}$, $N_\chi < \infty$ an othogonal direct sum with $T_\chi|_{H_i}$ irreducible and equivalent with $S_{2\pi}$.

Lemma A.4.3. Let $I(H^\chi, H^\chi)$ be the space of all bounded linear operators $A: H^\chi \to H^\chi$ so that $A \circ T_\chi(g) = T_\chi(g) \circ A$ for $g \in H_\beta$. Then

$$\dim I(H^\chi, H^\chi) = N_\chi^2 \quad .$$

Proof. Let

$$H^\chi = H_1 \oplus \cdots \oplus H_{N_\chi}$$

as in Theorem A.4.2. Let $P_i: H^\chi \to H_i$ be the orthogonal projection. Let $A \in I(H^\chi, H^\chi)$. Then

$$A = \sum_{i,j=1}^{N_\chi} P_i A P_j \quad .$$

$$P_i A P_j : H_j \to H_i \quad ,$$

$$(P_i \circ A \circ P_j) \circ T_\chi(g) = T_\chi(g) \circ P_i \circ A \circ P_j \quad .$$

By irreducibility of H_j and H_i $P_i \circ A \circ P_j = 0$ or $P_i \circ A \circ P_j \colon H_j \to H_i$ is bijective. Let $T_{ij} \colon H_j \to H_i$ be a unitary equivalence so that $T_{ij} = T_{ji}^{-1}$. Then $T_{ji} \circ P_i \circ A \circ P_j \colon H_j \to H_j$ is an intertwining operator. Thus

$$T_{ji} \circ P_i \circ A \circ P_j = \lambda_{ij} I \quad .$$

Thus

$$P_i \circ A \circ P_j = \lambda_{ij} T_{ij}$$

$$\lambda_{ij} \in \mathbb{C} \quad .$$

But then

$$A = \sum_{ij} \lambda_{ij} T_{ij} P_j$$

This clearly implies the result.

Let $\Gamma' \subset V$ be the set $\{v \in V \mid \beta(v,\gamma) \in \mathbb{Z}$ for all $\gamma \in \Gamma\}$. Let $e_1, \ldots, e_{2n}$ be a basis of Γ over $\mathbb{Z}$ so that

i) $\beta(e_i, e_j) = \beta(e_{n+i}, e_{n+j}) = 0, \quad i,j \leq n.$

ii) $\beta(e_i, e_{n+i}) = m_i, \quad m_i > 0$ and $m_1 | m_2 | \cdots | m_n$.
(See Lemma A.3.2.)

<u>Lemma A.4.4</u>. Γ' is a discrete subgroup of V containing Γ and

$$^{\#}(\Gamma'/\Gamma) = (m_1 \cdots m_n)^2 .$$

($A^{\#}$ is the number of elements in the set A.)

<u>Proof</u>. If $z \in \Gamma'$,

$$z = \sum_{i=1}^{n} z_i e_i + \sum_{i=1}^{n} w_i e_{n+1} .$$

$$\beta(z, e_i) = -w_i m_i , \qquad \beta(z, e_{n+1}) = m_i z_i$$

for $i = 1,\ldots,n$.

Thus $z_i = k_i/m_i$, $k_i \in \mathbb{Z}$, $w_i = \ell_i/m_i$, $\ell_i \in \mathbb{Z}$. Thus a basis of Γ' over $\mathbb{Z}$ is

$$\frac{1}{m_1} e_1, \ldots, \frac{1}{m_n} e_n, \frac{1}{m_1} e_{n+1}, \ldots, \frac{1}{m_n} e_{2n} .$$

Thus

$$\Gamma'/\Gamma = \mathop{\times}_{i=1}^{n} (\mathbb{Z}/\mathbb{Z}m_i \times \mathbb{Z}/\mathbb{Z}m_i) .$$

Q.E.D.

We show how one can define an element of $I(H^{\chi}, H^{\chi})$ for each $\gamma \in \Gamma'$. To do this, it is convenient to have another realization of H^{χ}.

If $f \in H^\chi$, define

$$B(f)(g) = g^{-1}\cdot f(g\cdot H_\beta(\Gamma)) \quad .$$

Then

$$B(f): H_\beta \to (L_\chi)_{eH_\beta(\Gamma)} \quad ,$$

$$e = 0 \quad .$$

Let $h \in H_\beta(\Gamma)$, $g \in H_\beta$. Then if $f \in L^\chi$,

$$\begin{aligned} B(f)(gh) &= (gh)^{-1} f(ghH_\beta(\Gamma)) \\ &= h^{-1}g^{-1}\cdot f(gH_\beta(\Gamma)) \\ &= \chi(h)^{-1}B(f)(g) \quad . \end{aligned}$$

Conversely, suppose

$$f: H \to (L_\chi)_{\overline{e}} \qquad (\overline{e} = e\cdot H_\beta(\Gamma)) \quad ,$$

f is, say, C^∞ and

I) $f(gh) = \chi(h)^{-1}f(g)$

Then, defining

$$\tilde{A}(f)(g) = g\cdot f(g)$$

we see that

$$\begin{aligned} \tilde{A}(f)(gh) &= ghf(gh) = g\chi(h)\chi(h)^{-1}f(g) \\ &= \tilde{A}(f)(g) \quad . \end{aligned}$$

Thus if we get $A(f)(gH_\beta(\Gamma)) = \tilde{A}(f)(g)$, then $A(f) \in C^\infty(T;L_\chi)$. Clearly,

II) $BA(f) = f$, $AB(f) = f$.

We identify $(L_\chi)_{\bar{e}}$ with $\mathbb{C}$. By $[0,z] \to z$. Let, for f_i satisfying I), $i = 1,2$

$$\langle f_1, f_2 \rangle = \int_{H_\beta/H_\beta(\Gamma) = T} f_1(g)\overline{f_2(g)}\, d(gH_\beta(\Gamma)) \quad .$$

Let $\tilde{H}^\chi$ be the completion of the space of functions satisfying I) relative to $\langle\ ,\ \rangle$. Then clearly

III) $A: \tilde{H}^\chi \to H^\chi$

$B: H^\chi \to \tilde{H}^\chi$

are unitary intertwining operators where

$$(\tilde{T}_\chi(x)f)(g) = f(x^{-1}g) \quad .$$

Let $\gamma \in \Gamma'$ and define for $f \in \tilde{H}^\chi = \tilde{H}^\chi \cap C^\infty(H_\beta)$,

$$(A_\gamma f)(g) = f(g(\gamma,0)) \quad .$$

We assert that

$$A_\gamma f \in \tilde{H}^\chi_\infty \quad .$$

Indeed, if $(\delta,t) \in H_\beta(\Gamma)$ then

$$
\begin{aligned}
(A_\gamma f)(g(\delta,t)) &= f(g(\delta,t)(\gamma,0)) \\
&= f(g(\gamma,0)(-\gamma,0)(\delta,t)(\gamma,0)) \\
&= f(g(\gamma,0)(-\gamma+\delta,\ t-\tfrac{1}{2}\beta(\gamma,\delta))(\gamma,0)) \\
&= f(g(\gamma,0)(\delta,t-\tfrac{1}{2}\beta(\gamma,\delta)+\tfrac{1}{2}\beta(-\gamma+\delta,\gamma)) \\
&= f(g(\gamma,0)(\delta,t-\beta(\gamma,\delta)) \\
&= \chi(\delta,t)^{-1}e^{2\pi i\beta(\gamma,\delta)}\ f(g\cdot(\gamma,0)) \\
&= \chi(\delta,t)^{-1}(A_\gamma f)(g)
\end{aligned}
$$

since $\beta(\gamma,\delta) \in \mathbb{Z}$ for all $\gamma \in L'$, $\delta \in L$.

Now the map $H_\beta/H_\beta(\Gamma) \to H_\beta/H_\beta(\Gamma)$ given by $g\cdot(H_\beta(\Gamma) \to g(\gamma,0)H_\beta(\Gamma)$ defines a diffeomorphism of $H_\beta/H_\beta(\Gamma)(T)$ for $\gamma \in \Gamma'$ given by translation by γ. Hence we see that $||A_\gamma f|| = ||f||$. Thus we have found for each $\gamma \in \Gamma'$,

$$A_\gamma\colon \tilde{H}^\chi \to \tilde{H}^\chi.$$

Clearly,

$$A_\gamma \circ \tilde{T}_\chi(g) = \tilde{T}_\chi(g)\circ A_\gamma$$

IV) If $\gamma,\gamma' \in \Gamma'$ then $A_\gamma \circ A_{\gamma'} = e^{-\pi i\beta(\gamma,\gamma')}A_{\gamma+\gamma'}$. Indeed

$$\begin{aligned}((A_\gamma \circ A_{\gamma'})f)(g) &= (A_{\gamma'}f)(g(\gamma,0)) \\ &= f(g(\gamma,0)(\gamma',0)) \\ &= f(g(\gamma+\gamma', \tfrac{1}{2}\beta(\gamma,\gamma')) \\ &= e^{-\pi i\beta(\gamma,\gamma')}\ f(g(\gamma+\gamma',0) \\ &= e^{-\pi i\beta(\gamma,\gamma')}\ A_{\gamma+\gamma'} \quad .\end{aligned}$$

V) If $\gamma \in \Gamma$ then $A_\gamma = e^{-\pi i F(\gamma)}\, I$.

This is obvious (see I).

Let $\gamma_1,\ldots,\gamma_{d^2}$ be a collection of representations for Γ'/Γ, $d = m_;,\ldots,m_n$. We assert that $A_{\gamma_1},\ldots,A_{\gamma_{d^2}}$ are linearly independent. Suppose

$$\sum a_i A_{\gamma_i} = 0 \quad .$$

Let $f \in \tilde{H}_\infty^X$ be such that

$$\text{supp } f\cdot(\gamma_j,0) \quad \text{supp } f = \emptyset$$

if $j \geq 2$. Then

$$0 = \sum a_i A_{\gamma_i} f = a_1 A_{\gamma_1} f \quad .$$

Suppose $f(\gamma_1,0) \neq 0$. Then $A_{\gamma_1} f \neq 0$ hence $a_1 = 0$. The result follows from the obvious induction.

We have proved

<u>Lemma A.4.5</u>. $N_\chi \geq m_1 \cdots m_n$.

Give $\tilde{H}^\chi_\infty$ the C^∞ topology. In this case it is just the subspace topology of $C^\infty(H_\beta;\mathbb{C})$. We will use the following result (whose proof is not hard).

<u>Proposition A.4.6</u>. Let $I^\infty(\tilde{H}^\chi,\tilde{H}^\chi)$ be the set of all $A: \tilde{H}^\chi_\infty \to \tilde{H}^\chi_\infty$ which are continuous in the C^∞ topology and

$$A\circ\tilde{T}_\chi(x) = \tilde{T}_\chi(x)\circ A$$

for $x \in H_\beta$.

Then

$$I^\infty(\tilde{H}^\chi,\tilde{H}^\chi) = I(\tilde{H}^\chi,\tilde{H}^\chi) \quad .$$

(For a proof, cf., G. Warner [28].)

Let $(\tilde{H}^\chi_\infty)'$ be the set of continuous linear functionals on $\tilde{H}^\chi_\infty$. If $A \in I^\infty(\tilde{H}^\chi,\tilde{H}^\chi)$, define

$$\delta_A(f) \to (Af)(0) \quad .$$

If $\delta_A = 0$ then

$$0 = \delta_A(\tilde{T}_\chi(g)f) = A(\tilde{T}_\chi(g)f)(0)$$

$$= (\tilde{T}_\chi(g)(Af))(0)$$

$$= (Af)(g^{-1}) \quad .$$

Thus $A = 0$. Now, if $h \in H_\beta(\Gamma)$ then

$$\begin{aligned}\delta_A(\tilde{T}_\chi(h)f) &= A(\tilde{T}_\chi(h)f)(0)\\ &= \tilde{T}_\chi(h(Af)(0)\\ &= A(f)(h^{-1})\\ &= \chi(h)A(f)(0)\\ &= \chi(h)\delta_A(f) \quad .\end{aligned}$$

Define

$$\tilde{T}'_\chi(g)\lambda = \lambda\circ\tilde{T}_\chi(g)^{-1}$$

for $\lambda \in (\tilde{H}^\chi_\infty)'$.

Let

$$i_\infty(\tilde{H}_\chi) = \{\lambda \in (\tilde{H}^\chi_\infty)' \mid \tilde{T}'_\chi(h)\lambda = \chi(h)^{-1}\lambda \quad \text{for} \quad h \in H_\beta(\Gamma)\} \ .$$

Lemma A.4.7. The map $A \mapsto \delta_A$ of $I^\infty(\tilde{H}_\chi, \tilde{H}_\chi)$ to $i_\infty(\tilde{H}_\chi)$ is a bijection.

Proof. The map is clearly injective. If $\lambda \in i_\infty(\tilde{H}_\chi)$, define

$$A_\lambda(f)(g) = \lambda(\tilde{T}_\chi(g)^{-1}f) \quad .$$

Then

$$\begin{aligned}(A_\lambda(f))(gh) &= \lambda(\tilde{T}_\chi(h)^{-1}\tilde{T}_\chi(g)^{-1}f) \\ &= \tilde{T}'_\chi(h)\lambda(\tilde{T}_\chi(g)^{-1}f) \\ &= \chi(h)^{-1}A_\lambda(f)(g) \quad .\end{aligned}$$

Hence,

$$A_\lambda : \tilde{H}^\chi_\infty \to \tilde{H}^\chi_\infty \quad .$$

Q.E.D.

If $\phi \in C_c^\infty(V)$ define

$$f_\phi((v,t)) = e^{-2\pi it} \sum_{\gamma\in\Gamma} e^{\pi iF(\gamma)} e^{-\pi i\beta(v,\gamma)} \phi(v+\gamma) \quad .$$

Then $f_\phi \in \tilde{H}^\chi_\infty$. It is easy to see that if $\lambda \in i_\infty(\tilde{H}^\chi)$ and $\lambda(f_\phi) = 0$ for all $\phi \in C_c^\infty(V)$ then $\lambda = 0$.

Let $\hat{\lambda}(\phi) = \lambda(f_\phi)$. Then $\hat{\lambda}$ satisfies

1) $\hat{\lambda}(t_\gamma\phi) = e^{-\pi iF(\gamma)} e^{\pi i\beta(\cdot,\gamma)} \hat{\lambda}(\phi)$

2) $\hat{\lambda}(t_\gamma\phi) = e^{-\pi iF(\gamma)} e^{-\pi i\beta(\cdot,\gamma)} \hat{\lambda}(\phi)$.

(If $\gamma \in \Gamma$, $(t_\gamma\phi)(x) = \phi(x+\gamma)$.) Hence we see

3) $e^{2\pi i\beta(\cdot,\gamma)} \hat{\lambda} = \hat{\lambda}$.

Thus

$$\operatorname{supp} \hat{\lambda} \subset \Gamma' \quad .$$

Let for $v \in V$, $\delta_v(\phi) = \phi(v)$. Then

$$\hat{\lambda} = \sum_{\gamma \in \Gamma'} C(\gamma)(D_\gamma \delta_\gamma)$$

where D_γ is a differential operator. We assert that D_γ is of order 0.

In fact, if $\gamma \in \Gamma$ then

$$\hat{\lambda} \circ t_\gamma = \sum_{\gamma' \in \Gamma'} C(\gamma')(D_{\gamma'} \delta_{\gamma'}) \circ t_\gamma$$

$$= e^{-\pi i F(\gamma)} e^{-\pi i \beta(\cdot, \gamma)} \hat{\lambda} .$$

Hence

$$C(\gamma')(D_{\gamma'} \delta_{\gamma'}) \circ t_\gamma = e^{+\pi i F(\gamma)} e^{\pi i \beta(\gamma', \gamma)} C(\gamma' - \gamma) D_{\gamma' - \gamma} \delta_{\gamma'}$$

We therefore see that

$$\hat{\lambda} = \sum_{i=1}^{d^2} C_i \sum_{\gamma \in \Gamma} e^{\pi i F(\gamma)} e^{\pi i \beta(\gamma_i, \gamma)} D_i \delta_{\gamma_i - \gamma}$$

with D_i a differential operator.

Let $\lambda_\gamma = \lambda_{A_\gamma}$. Then an easy computation gives

$$\hat{\lambda}_{\gamma_i} = \sum_{\gamma \in \Gamma} e^{\pi i F(\gamma)} e^{\pi i \beta(\gamma_i, \gamma)} \delta_{\gamma_i - \gamma}$$

Now the intertwining operator corresponding to $D_i \hat{\lambda}_{\gamma_i}$ is of the form

$$A_{\gamma_i} R_u f \quad .$$

R_u differentiation on the right by an element, u, of the universal enveloping algebra of h_β.

But then $Ad(z)u = u$ for $z \in H_\beta(\Gamma)$.

<u>Lemma A.4.8</u>. If (ρ, V) is a finite dimensional representation of H_β and $v \in V$, $\sigma \neq 0$ is such that

$$\rho(H_\beta(\Gamma))v = v \quad .$$

Then

$$v = \sum v_j$$

and

$$\rho(h)v_j = \chi_j(h)v_j$$

$$\text{for } h \in H_\beta$$

where χ_j is a unitary character of H_β.

<u>Proof</u>. Let $V_1 = \{\omega \in V \mid \overline{\rho(H_\beta)v} \text{ is compact}\}$. Then $\rho(h)V_1 \subset V_1$ for $h \in H_\beta$. Clearly V_1 is a subspace of V. Clearly $v \in V_1$.

Let $e_1, \ldots, e_p$ be a basis of V_1. Then

$$\rho(h)e_i = \sum_{j=1}^{n} \tilde{\rho}(h)_{ji} e_j$$

and

$$\tilde{\rho}: H_\beta \to GL(p,\mathbb{C}) \quad .$$

Clearly $\overline{\tilde{\rho}(H_\beta)}$ is compact. Hence $\overline{\tilde{\rho}(H_\beta)}$ is a compact subgroup of $GL(p,\mathbb{C})$. We can thus integrate any inner product on V_1 to make the representation (ρ,V_1) of H_β unitary. But then

$$V_1 = \sum V_{ij}$$

with

$$\rho(h)|_{V_{ij}} = \chi_j(h)I \quad ,$$

χ_j a unitary character (the only irreducible finite dimensional unitary representation of H_β are characters). Q.E.D.

Using Lemma A.4.8 we conclude that

$$Ad(z)u = u$$

$$\text{for} \quad z \in H_\beta \quad .$$

Indeed, u is contained in a finite dimensional invariant subspace V of the universal enveloping algebra of h_β. This space is equivalent with a quotient representation of

$$\sum_{j=1}^{\delta} \otimes^j h_\beta$$

where H_β acts on h_β be the adjoint representation. But the action of H_β on h_β can be put in upper triangular form with ones along the diagonal. Hence, the only unitary characters that can appear in V are the trivial character. But, now

$$R_u \circ \tilde{T}_\chi(x) = \tilde{T}_\chi(x) \circ R_u$$
$$\text{for } x \in H_\beta \quad .$$

From this and the fact that $\tilde{H}^\chi$ is a finite direct sum of representations equivalent with $S_{2\pi}$ we find that $R_u f = cf$, $c \in \mathbb{C}$ for $f \in \tilde{H}^\chi_\infty$. Thus

$$A_{\gamma_i} R_u = cA_{\gamma_i} \quad .$$

We have proved

<u>Theorem A.4.9</u>. $N_\chi = m_1 \cdots m_n$.

Now we return to the notation at the end of A.3. That is, we assume

a) V is actually complex n-dimensional with complex structure J.

b) $B = \operatorname{Im} \langle \, , \rangle$ where $\langle \, , \rangle$ is a positive definite Hermitian inner product on V.

We rephrase the results of A.3 in the following form

Theorem A.4.10. Let

$$\Theta_F = \{\theta : V \to \mathbb{C} \mid \theta \text{ holomorphic and}$$
$$\theta(v+\gamma) = \theta(v)\exp(\pi(-\langle\gamma,\gamma/2\rangle + \langle v,\gamma\rangle - iF(\gamma))\}$$

Then the map $f \mapsto \theta_f$ of

$$U_F = \{f \in \tilde{H}_\infty^\chi \mid \tilde{T}_\chi(V^-,0)\cdot f = 0\}$$

given by $\theta_f(v) = f(p^+(v),0)$ (here f is looked upon as a cross-section of L_χ) is a bijection. We note that if $f \in \tilde{H}_\infty^\chi$ then

$$\tilde{T}_\chi(X)f = \frac{d}{dt}\Big|_{t=0} \tilde{T}_\chi(\exp tX)f \quad ,$$

for $\chi \in h_\beta$

$$\tilde{T}_\chi(X + iY) = \tilde{T}_\chi(X) + i\tilde{T}_\chi(Y) \quad .$$

Using the above theorem we prove

Theorem A.4.11 (Frobenius). $\dim \Theta_F = m_1 \cdots m_n$.

Proof. We need only prove that

$$\dim\{f \in \mathcal{S}(\mathbb{R}^n) \mid S_{2\pi}(X)f = 0 \text{ for } X \in (V^-,0)\} = 1 \quad .$$

By a change of scale we can take 2π to be 1. We can choose a basis $e_1,\ldots,e_{2n}$ of V so that $Je_i = -e_{n+i}$, $Je_{n+i} = +e_i$,

$i = 1,\dots,n$ and $\beta(e_i,e_j) = \beta(e_{n+i},e_{n+j}) = 0$, $\beta(e_i,e_{n+i}) = 1$. Then

$$S_1(e_i,0)f = ix_i f \quad ,$$

$$S_1(e_{n+i},0)f = -\frac{\partial}{\partial x_i} f \quad .$$

We are looking at elements of the form $e_i + iJe_i$, $i = 1,\dots,n$ which go over to operators

$$ix_i + i\frac{\partial}{\partial x_i} = +i\left(\frac{\partial}{\partial x_i} + x_i\right) \quad .$$

Now if

$$\frac{\partial}{\partial x_i} f = -x_i f \quad ,$$

$i = 1,\dots,n$ then

$$f(X) = c\, e^{-\|X\|^2/2} \quad .$$

Thus, the result.

Chapter 5

THE METAPLECTIC REPRESENTATION

In this chapter we study an important unitary representation of the two fold covering of the symplectic group, the metaplectic representation. The existence of this representation is predicted by the quantization rules of the early quantum mechanics. However, the rigorous proof of its existence did not appear until the 1950's due to work of Segal, Shale [22] and somewhat later, Weil [31]. It was Shale [22] who observed that the metaplectic representation should be looked upon as an analogue of the spin representation of the orthogonal groups (symmetric tensors taking the place of anti-symmetric tensors). In our development we follow this philosophy quite closely. The reader who knows the theory of the spin representation will recognize that the steps involved in our development of the metaplectic representation are completely analogous to the steps involved in the construction of the spin representation. A very similar development can be found in Igusa [11].

In the last section of this chapter we show how the Hermite functions give the Fourier series expansion relative to a certain circle in the metaplectic representation. Our development of the Hermite functions is certainly not the easiest. However, it is canonical in the sense that the "tricks" in the usual development became "natural" in our group theoretic context.

1. THE METAPLECTIC GROUP

Let H_n be the $2n+1$ dimensional Heisenberg group. That is, if

$$\beta\left(\begin{bmatrix} x \\ y \end{bmatrix}, \begin{bmatrix} x' \\ y' \end{bmatrix}\right) = \langle x,y'\rangle - \langle y,x'\rangle$$

($x,x',y,y' \in \mathbb{R}^n$ and $\langle\ ,\ \rangle$ is the standard inner product on $\mathbb{R}^n$) then H_n is $\mathbb{R}^{2n} \times \mathbb{R}$ with

$$(x,t)(u,s) = (x+u,\ t+s+\tfrac{1}{2}\beta(x,u)) \quad .$$

In Chapter 4, Section 4 we gave a classification of the irreducible unitary representation of H_n. In this section we look at a particular irreducible unitary representation $(\pi, L^2(\mathbb{R}^n))$, $\pi = S_1$. Recall that

$$\left(S_1\left(\begin{bmatrix} x \\ y \end{bmatrix},\ t\right)f\right)(z) = e^{i(t+\langle x, z-\frac{1}{2}y\rangle)}\ f(z-y) \quad ,$$

$$x,y,z \in \mathbb{R}^n, \quad t \in \mathbb{R} \quad .$$

Let $F\colon L^2(\mathbb{R}^n) \to L^2(\mathbb{R}^n)$ be the Fourier transform. A direct calculation yields the following result.

<u>Lemma 1.1</u>. $\quad F\pi\left(\begin{bmatrix} x \\ y \end{bmatrix},\ t\right)F^{-1} = \pi\left(\begin{bmatrix} -y \\ x \end{bmatrix},\ t\right)$

Let $U(L^2(\mathbb{R}^n))$ be the unitary group of $L^2(\mathbb{R}^n)$, equipped with the <u>strong</u> topology. (See Sections 4.3 or 4.4.)

<u>Lemma 1.2</u>. Let $A = \pi(H_n) \subset U(L^2(\mathbb{R}^n))$ with the subspace topology. Then $\pi\colon H_n \to A$ is a homeomorphism.

<u>Proof</u>. We have already seen that π is continuous. We must therefore show that if $\pi(h)$ and $\pi(h')$ are close to each other in A then they are close to each other in H_n.

If $\Phi \in L^2(\mathbb{R}^n)$, $||\Phi|| = 1$ and if

$$\left|\left|\pi\left(\begin{bmatrix} x \\ y \end{bmatrix}, t\right)\Phi - \Phi\right|\right| < \sqrt{2} \quad ,$$

then

$$2 > 2||\Phi||^2 + 2 \operatorname{Re} \int_{\mathbb{R}^n} e^{i(t - <x, z - \frac{1}{2}y>)} \Phi(z - y)\Phi(z)\, dz$$

Thus $\Phi(z - y)\Phi(z) \not\equiv 0$. Suppose that $\Phi \in C_c^\infty(\mathbb{R}^n)$ and $\operatorname{supp} \Phi \subset \{z \in \mathbb{R}^n \mid ||z|| < \epsilon\} = B_\epsilon(0)$. Then there is $z \in B(0)$ so that $||z-y|| < \epsilon$. Hence $||y|| < 2\epsilon$. Using $F^{-1}\Phi$ in the above argument we find $||x|| < 2\epsilon$. Also

$$\pi\left(\begin{bmatrix} x \\ y \end{bmatrix}, t\right)\Phi = e^{it}\, \pi\left(\begin{bmatrix} x \\ y \end{bmatrix}, 0\right)\Phi \quad .$$

It is therefore clear that with appropriate choice of Φ we must have $|t| < 2\epsilon$.

The result now follows by observing that

$$||\pi(h)\phi - \pi(h')\phi|| = ||\pi(h'^{-1}h)\Phi - \Phi|| \quad .$$

Q.E.D.

Recall that $Sp(n,\mathbb{R})$ (the symplectic group of $\mathbb{R}^{2n}$) is the subgroup of those $g \in GL(2n,\mathbb{R})$ such that

$$\beta(g\cdot u,\ g\cdot w) = \beta(u,w)$$

for all $u,w \in \mathbb{R}^{2n}$.

Let G denote the subgroup of $U(L^2(\mathbb{R}^n))$ consisting of all $g \in U(L^2(\mathbb{R}^n))$ such that for each $u \in \mathbb{R}^{2n}$, $t \in \mathbb{R}$ there is $u' \in \mathbb{R}^{2n}$ so that

$$g\pi(u,t)g^{-1} = \pi(u',t) \quad .$$

If $g \in G$ then Lemma 1.2 implies that

$$g\pi(u,t)g^{-1} = \pi(\nu(g)(u),t)$$

with $\nu(g)\colon \mathbb{R}^{2n} \to \mathbb{R}^{2n}$ a homeomorphism.

Lemma 1.3. 1) If $g \in G$ then $\nu(g) \in Sp(n,\mathbb{R})$.

2) $\nu\colon G \to Sp(n,\mathbb{R})$ is continuous (here $G \subset U(L^2(\mathbb{R}^n))$ has the subspace topology).

3) $\operatorname{Ker}\nu = \{\xi I \mid |\xi| = 1\}$.

Proof.

$$\begin{aligned} g\pi(z,0)\pi(w,0)g^{-1} &= g\pi(z,0)g^{-1}g\pi(w,0)g^{-1} \\ &= \pi(\nu(g)(z),0)\pi(\nu(g)(w),0) \\ &= \pi(\nu(g)(z)+\nu(g)(w),\tfrac{1}{2}\beta(\nu(g)(z),\nu(g)(w)) \end{aligned}$$

Also,

$$g\pi(z,0)\pi(w,0)g^{-1} = g\pi(z+w,\tfrac{1}{2}\beta(z,w))g^{-1}$$

$$= \pi(\nu(g)(z+w),\tfrac{1}{2}\beta(z,w)) \quad .$$

Hence,

a) $\nu(g)(z) + \nu(g)(w) = \nu(g)(z+w)$

and

b) $\beta(\nu(g)(z),\nu(g)(w)) = \beta(z,w) \quad .$

Since $\nu(g)$ is continuous we see that a) implies that $\nu(g)$ is linear. b) implies $\nu(g) \in Sp(n,\mathbb{R})$. The continutiy follows from Lemma 1.2. This proves 1) and 2).

If $\nu(g) = I$, then $g\pi(z,t)g^{-1} = \pi(z,t)$ for all $z \in \mathbb{R}^{2n}$, $t \in \mathbb{R}$. Hence $g = \lambda I$ since $(\pi, L^2(\mathbb{R}^n))$ is irreducible. This clearly implies 3).

<u>Definition 1.4</u>. The <u>metaplectic group</u>, $Mp(n,\mathbb{R})$, is the closure of the subgroup of G generated by all commutators (elements of the form $ghg^{-1}h^{-1}$, $g,h \in G$).

In the next section we will study the structure of $Sp(n,\mathbb{R})$. We will then use those results to study the structure of $Mp(n,\mathbb{R})$ and the map $\nu\colon Mp(n,\mathbb{R}) \to Sp(n,\mathbb{R})$.

2. THE SYMPLECTIC GROUP

Recall that $Sp(n,\mathbb{R})$, the symplectic group, is the group of all $g \in GL(2n,\mathbb{R})$ so that

$$\beta(gu,gw) = \beta(u,w)$$

for all $u,w \in \mathbb{R}^{2n}$.

We note that if

$$\langle u,w\rangle = \sum_{i=1}^{2n} u_i w_i$$

and

$$J = \begin{bmatrix} 0 & I_n \\ -I_n & 0 \end{bmatrix} ,$$

I_n the $n \times n$ identity matrix, then

$$\beta(u,w) = \langle u,Jw\rangle .$$

Hence

$$Sp(n,\mathbb{R}) = \{g \in GL(2n,\mathbb{R}) \mid gJ^tg = J\} .$$

In particular, if $g \in Sp(n,\mathbb{R})$, then ${}^tg \in Sp(n,\mathbb{R})$.

Lemma 2.1. Let N be the subgroup of $GL(2n,\mathbb{R})$ consisting of the matrices of the form

$$\begin{bmatrix} I_n & X \\ 0 & I_n \end{bmatrix}$$

$X \in M_n(\mathbb{R})$, ${}^tX = X$. Let M be the subgroup of $GL(2n,\mathbb{R})$ consisting of the matrices of the form

$$\left[\begin{array}{c|c} A & 0 \\ \hline 0 & {}^tA^{-1} \end{array}\right]$$

with $A \in GL(2,\mathbb{R})$. Then M and N are subgroups of $Sp(n,\mathbb{R})$. Furthermore, every element of $Sp(n,\mathbb{R})$ is a product of elements of M, elements of N and some J's.

Note. In the proof of this lemma we will develop many structural results for $Sp(n,\mathbb{R})$ which will be used later.

Proof. Let $\mathfrak{g} \subset M_{2n}(\mathbb{R})$ be the Lie algebra of $Sp(n,\mathbb{R})$. By differentiating the definition of $Sp(n,\mathbb{R})$ we see

i) $\mathfrak{g} = \{X \in M_{2n}(\mathbb{R}) \mid XJ + J\,{}^tX = 0\}$

Set $\mathfrak{p} = \{X \in \mathfrak{g} \mid {}^tX = X\}$.

ii) $Sp(n,\mathbb{R}) = (Sp(n,\mathbb{R}) \cap O(2n)) \cdot \exp(\mathfrak{p})$.

To prove ii) we note that if $g \in Sp(n,\mathbb{R})$, then, in particular, $g \in GL(2n,\mathbb{R})$. Hence $g = UA$ with $U \in O(2n)$, $A \in GL(2n,\mathbb{R})$, ${}^tA = A$ and A is positive definite. Now

$${}^{t}gg = A^2 \in Sp(n,\mathbb{R}) .$$

Also $A^2 = \exp(Y)$, $Y \in M_{2n}(\mathbb{R})$, ${}^{t}Y = Y$. Hence,

$$\exp(kY)J \exp(kY) = J$$

$$\text{for all } k \in \mathbb{Z} .$$

This implies (cf., Wallach [26], 3.2.7) that

$$\exp(tY)J \exp(-tY) = J$$

$$\text{for all } t \in \mathbb{R} .$$

Hence

$$YJ + JY = 0 .$$

This says $Y \in p$. This clearly implies that $U \in O(2n) \cap Sp(n,\mathbb{R})$. We have proven ii).

If $g \in O(2n) \cap Sp(n,\mathbb{R})$, then $gJ{}^{t}g = J$. But ${}^{t}g = g^{-1}$. Hence $gJ = Jg$. If we look upon $(\mathbb{R}^{2n}, J)$ as $\mathbb{C}^n$ $(J^2 = -I)$ and if $g \in O(2n) \cap Sp(n,\mathbb{R})$, then g is complex linear.

Define for $x,y \in \mathbb{R}^{2n}$,

$$(x,y) = \langle x,y\rangle + i\beta(x,y) .$$

Then $(\, , \,)$ defines a positive definite Hermitian inner product on $\mathbb{C}^n$. It is easily seen that $U(n)$ is then $O(2n) \cap Sp(n,\mathbb{R})$. We have shown

iii) $O(2n) \cap Sp(n,\mathbb{R})$ is Lie isomorphic with $U(n)$.

In particular, we have

iv) $Sp(n,\mathbb{R})$ is connected.

Indeed, $U(n)$ is connected, $\exp(p)$ is connected. The result now follows from ii).

v) Set

$$\overline{N} = \left\{ \begin{bmatrix} I & 0 \\ X & I \end{bmatrix} \middle| \; {}^tX = X,\ X \in M_n(\mathbb{R}) \right\}$$

Then $\overline{N} = JNJ$. Also

$$\overline{N}MN = \left\{ \begin{bmatrix} A & B \\ C & D \end{bmatrix} \in Sp(n,\mathbb{R}) \quad A,B,C,D \in M_n(\mathbb{R}) \text{ and } \det A \neq 0 \right\} .$$

The first assertion of v) is obvious. To prove the second assertion we first note that if $X,Y \in M_n(\mathbb{R})$, ${}^tX = X$, ${}^tY = Y$, and if $E \in GL(n,\mathbb{R})$, then

$$g = \begin{bmatrix} I & 0 \\ X & I \end{bmatrix} \begin{bmatrix} E & 0 \\ 0 & {}^tE^{-1} \end{bmatrix} \begin{bmatrix} I & Y \\ 0 & I \end{bmatrix} = \begin{bmatrix} E & EY \\ XE & XEY + {}^tE^{-1} \end{bmatrix}$$

Thus if

$$g = \begin{bmatrix} A & B \\ C & D \end{bmatrix} \in Sp(n,\mathbb{R})$$

and $\det A \neq 0$, set

$$E = A, \quad Y = A^{-1}B, \quad X = CA^{-1} \quad .$$

We must check

a) ${}^tY = Y, \quad {}^tX = X$

b) $D = CA^{-1}B + {}^tA^{-1}$.

a), b) are fairly easy consequences of $gJ{}^tg = J$ and ${}^tgJg = J$ (write these identities out in terms of A,B,C,D). We have thus proved v).

We can now prove the lemma. Let G_1 be the subgroup of $Sp(n,\mathbb{R})$ generated by J, N and M. Then by v),

$$G_1 \supset \left\{ \begin{bmatrix} A & B \\ C & D \end{bmatrix} \varepsilon \; Sp(n,\mathbb{R}) \mid \det A \neq 0 \right\} = V \quad .$$

Since V is open in $Sp(n,\mathbb{R})$, G_1 is an open subgroup of $Sp(n,\mathbb{R})$. Hence G_1 is closed (indeed, $Sp(n,\mathbb{R}) - G_1 = \cup_{g \notin G_1} gG_1$). Since $Sp(n,\mathbb{R})$ is connected ((iv) above), $G_1 = Sp(n,\mathbb{R})$. Q.E.D.

3. THE STRUCTURE OF THE METAPLECTIC GROUP

We return to the notation of Section 1. In particular, $G \subset U(L^2(\mathbb{R}^n))$ is the subgroup of all $g \;\varepsilon\; U(L^2(\mathbb{R}^n))$ so that

$$g\pi(u,t)g^{-1} = \pi(u',t) \quad ,$$

$$u \;\varepsilon\; \mathbb{R}^{2n} \quad .$$

We give G the subspace topology. We have seen $\nu: G \to Sp(n,\mathbb{R})$ is a continuous homomorphism with

$$\text{Ker } \nu = \{\xi I \mid |\xi| = 1\} .$$

<u>Lemma 3.1</u>. G has the structure of a Lie group. $\nu: G \to Sp(n,\mathbb{R})$ is surjective and is a Lie homomorphism.

<u>Proof</u>. We need only prove the first two assertions. The last follows from the fact that a continuous homomorphism of Lie groups is a Lie homomorphism. We will only sketch the proof of this result leaving details to the reader. We note that the first statement is a special case of the theorem that an extension of a Lie group by a Lie group is a Lie group. We use the notation (and results of the proof of Lemma 2.1).

Set

$$V = \left\{ g = \begin{bmatrix} A & B \\ C & D \end{bmatrix} \middle| \; g \in Sp(n,\mathbb{R}), \;\; A \in GL(n,\mathbb{R}) \right\} .$$

If $g \in V$, then

$$g = \bar{n}(g)m(g)n(g)$$

with

1)
$$\bar{n}(g) = \begin{bmatrix} I & 0 \\ A^{-1}B & I \end{bmatrix}, \quad m(g) = \begin{bmatrix} A & 0 \\ 0 & {}^tA^{-1} \end{bmatrix},$$
$$n(g) = \begin{bmatrix} I & CA^{-1} \\ 0 & I \end{bmatrix}$$

Clearly, $\bar{n}$, m and n are C^∞ functions on V.

We now introduce some notation that will be used throughout this section. If $A \in M_n(\mathbb{R})$ and ${}^tA = A$, define

$$(\mu(A)f)(z) = e^{i<Az,z>/2} f(z)$$

for $f \in L^2(\mathbb{R}^n)$.

2) $\mu(A) \in G$ and $\nu(\mu(A)) = \begin{bmatrix} I & A \\ 0 & I \end{bmatrix}$

This is a straightforward computation.

3) $\mathcal{F} \in G$ and $\nu(\mathcal{F}) = -J$.

This follows from Lemma 1.1.

Set for $A \in GL(n,\mathbb{R})$

$$(\alpha(A)f)(z) = |\det A|^{1/2} f({}^tAz)$$

for $f \in L^2(\mathbb{R}^n)$.

4) $\alpha(A) \in G$ and $\nu(\alpha(A)) = \begin{bmatrix} A & 0 \\ 0 & {}^tA^{-1} \end{bmatrix} \in Sp(n,\mathbb{R})$

The change of variables formula implies that

$$\alpha(A) \in U(L^2(\mathbb{R}^n)) .$$

A direct computation shows that

$$\left(\alpha(A)\pi\left(\begin{bmatrix}x\\y\end{bmatrix},t\right)\alpha(A)^{-1}f\right)(z) = \exp\left(i(t+\langle x,{}^tAz-\tfrac{1}{2}y\rangle\right)f(z-{}^tA^{-1}y)$$

$$= \exp\left(i(t+\langle Ax,z-\tfrac{1}{2}{}^tA^{-1}y\rangle\right)f(z-{}^tA^{-1}y)$$

$$= \pi\left(\left(\begin{bmatrix}Ax\\{}^tA^{-1}y\end{bmatrix},t\right)f\right)(z) \quad .$$

This proves 4).

Define $\psi: V \to G$ by

$$\psi(g) = \mathcal{F}\mu(-AB^{-1})\mathcal{F}^{-1}\alpha(A)\mu(CA^{-1})$$

(see 1)). Then

$$\nu(\psi(g)) = g$$

for $g \in V$.

Clearly, $\psi: V \to \psi(V)$ is a homeomorphism. Set

$$\Phi(t,g) = t\psi(g)$$

for $t \in T'$ and $g \in V$.

It is easily checked that $\Phi: T' \times V \to \Phi(T' \times V)$ is a homeomorphism. Give G a manifold structure by taking the manifold structure on $\Phi(t' \times V)$ and "translating around" these sets. ($\Phi(T' \times V)$ is open in G by Lemma 1.3.)

Lemma 2.1 now implies that ν is surjective. Q.E.D.

By definition, $Mp(n,\mathbb{R})$ is the closure of the group generated by the commutators in G. Hence $Mp(n,\mathbb{R})$ inherits a Lie group structure from G.

Theorem 3.2. $Mp(n,\mathbb{R})$ is the closure of the subgroup of $U(L^2(\mathbb{R}^n))$ generated by the elements $\mu(A)$ and $\mathcal{F}\mu(A)\mathcal{F}^{-1}$, ${}^tA = A$ (see 2) in the proof of Lemma 3.1). Furthermore,

$$\nu(Mp(n,\mathbb{R})) = Sp(n,\mathbb{R})$$

and

$$\operatorname{Ker} \nu|_{Mp(n,\mathbb{R})} = \pm I .$$

Proof. We first note that if $X \in M_n(\mathbb{R})$, ${}^tX = X$ and $A \in GL(n,\mathbb{R})$, then

$$\alpha(A)\mu(X)\alpha(A)^{-1}\mu(X)^{-1} = \mu(AX{}^tA - X) .$$

(See 4) in the proof of Lemma 3.1 for the definition of $\alpha(A)$.) Hence, if $A = \sqrt{2}\, I$ we see that

$$\mu(X) = \alpha(A)\mu(X)\alpha(A)^{-1}\mu(X)^{-1} .$$

Hence $\mu(X) \in Mp(n,\mathbb{R})$. Since $Mp(n,\mathbb{R})$ is clearly normal in G, $\mathcal{F}\mu(X)\mathcal{F}^{-1} \in Mp(n,\mathbb{R})$.

1) The Lie algebra of $Sp(n,\mathbb{R})$ is generated by the subalgebras

$$n = \left\{ \begin{bmatrix} 0 & X \\ 0 & 0 \end{bmatrix} \middle| \; X \in M_n(\mathbb{R}), \; {}^tX = X \right\}$$

$$\bar{n} = \left\{ \begin{bmatrix} 0 & 0 \\ X & 0 \end{bmatrix} \middle| \; X \in M_n(\mathbb{R}), \; {}^tX = X \right\}$$

This is an easy exercise since the Lie algebra of $Sp(n,\mathbb{R})$, $sp(n,\mathbb{R})$, is equal to $\bar{n} \oplus M \oplus n$ with

$$M = \left\{ \left[\begin{array}{c|c} X & 0 \\ \hline 0 & -{}^tX \end{array} \right] \middle| \; X \in M_n(\mathbb{R}) \right\}$$

1) immediately implies that if $G_0 \subset Mp(n,\mathbb{R})$ is the closure of the subgroup generated by the $\mu(X)$ and $F\mu(X)F^{-1}$, $X \in M_n(\mathbb{R})$, ${}^tX = X$, then $\nu\colon G_0 \to Sp(n,\mathbb{R})$ is surjective.

Now it is clear that G, G_0, and $Mp(n,\mathbb{R})$ are connected. Let g be the Lie algebra of G. Then, since $\nu_*\colon g \to sp(n,\mathbb{R})$ is a surjective Lie algebra homomorphism with $\operatorname{Ker} \nu_* \subset$ center of g, we see that $\nu_*([g,g]) = sp(n,\mathbb{R})$ and $\nu_*\colon [g,g] \to sp(n,\mathbb{R})$ is a Lie algebra isomorphism. Here we use the fact that since $sp(n,\mathbb{R})$ is semi-simple, $g = z \oplus [g,g]$, z the center.

But then G_0 and $Mp(n,\mathbb{R})$ have the same Lie algebra which is isomorphic with $Sp(n,\mathbb{R})$. This implies $G_0 = Mp(n,\mathbb{R})$ and $\nu\colon Mp(n,\mathbb{R}) \to Sp(n,\mathbb{R})$ is a covering map.

We also note $\operatorname{Ker} \nu|_{Mp(n,\mathbb{R})}$ is discrete and contained in $\{\lambda I \big| \, |\lambda| = 1\}$. Hence $\operatorname{Ker} \nu|_{Mp(n,\mathbb{R})}$ is finite. To complete the proof of the lemma we need only show that $\operatorname{Ker} \nu|_{Mp(n,\mathbb{R})} = \pm I$.

Let

$$K = O(2n) \cap Sp(n,\mathbb{R})$$

and

$$\tilde{K} = \nu^{-1}(K) \cap Mp(n,\mathbb{R}) \quad .$$

Then $Sp(n,\mathbb{R})/K$ is simply connected (see ii), the proof of Lemma 2.1). Hence, since

$$\eta\colon Mp(n,\mathbb{R})/\tilde{K} \to Sp(n,\mathbb{R})/K \quad ,$$

$\eta(g\tilde{K}) = \nu(g)K$ is a covering map, $\tilde{K}$ is connected, and

$$\operatorname{Ker}\left[\nu \Big|_{Mp(n,\mathbb{R})}\right] \subset \tilde{K} \quad .$$

Let T be the maximal torus in K of elements of the form

$$\left[\begin{array}{ccc|ccc} \cos\theta_1 & & 0 & \sin\theta_1 & & 0 \\ & \ddots & & & \ddots & \\ 0 & & \cos\theta_n & 0 & & \sin\theta_n \\ \hline -\sin\theta_1 & & 0 & \cos\theta_1 & & 0 \\ & \ddots & & & \ddots & \\ 0 & & -\sin\theta_n & 0 & & \cos\theta_n \end{array}\right] = t(\theta_1,\ldots,\theta_n)$$

Then $\tilde{T} = \nu^{-1}(T) \cap \tilde{K}$ is a maximal torus of $\tilde{K}$ and

$$\operatorname{Ker} \nu|_{Mp(n,\mathbb{R})} \subset \tilde{T} \quad .$$

If

$$\prod_{i=1}^{n} \cos\theta_i \neq 0$$

then

$$t(\theta_1,\dots,\theta_n) = \left[\begin{array}{c|c} I & 0 \\ \hline \begin{matrix} -\tan\theta_1 & & \\ & \ddots & \\ & & -\tan\theta_n \end{matrix} & I \end{array}\right]$$

$$\cdot \left[\begin{array}{c|c} \begin{matrix} \cos\theta_1 & & 0 \\ & \ddots & \\ 0 & & \cos\theta_n \end{matrix} & 0 \\ \hline 0 & \begin{matrix} (\cos\theta_1)^{-1} & & 0 \\ & \ddots & \\ 0 & & (\cos\theta_n)^{-1} \end{matrix} \end{array}\right]$$

$$\cdot \left[\begin{array}{cc} I & \begin{matrix} \tan\theta_1 & & 0 \\ & \ddots & \\ 0 & & \tan\theta_n \end{matrix} \\ 0 & I \end{array}\right]$$

Now this says,

$$t(\theta_1,\dots,\theta_n) = \nu\left(\mathcal{F}\mu\begin{bmatrix}\tan\theta_1 & & 0\\ & \ddots & \\ 0 & & \tan\theta_n\end{bmatrix}\mathcal{F}^{-1}\alpha\left(\begin{bmatrix}\cos\theta_1 & & 0\\ & \ddots & \\ 0 & & \cos\theta_n\end{bmatrix}\right)\right.$$

$$\left.\cdot\,\mu\left(\begin{bmatrix}\tan\theta_1 & & 0\\ & \ddots & \\ 0 & & \tan\theta_n\end{bmatrix}\right)\right)$$

2) Let $g(x) = e^{-\|x\|^2/2}$, then

$$\mathcal{F}\mu\begin{bmatrix}\tan\theta_1 & & 0\\ & \ddots & \\ 0 & & \tan\theta_n\end{bmatrix}\mathcal{F}^{-1}\alpha\begin{bmatrix}\cos\theta_1 & & 0\\ & \ddots & \\ 0 & & \cos\theta_n\end{bmatrix}\mu\begin{bmatrix}\tan\theta_1 & & 0\\ & \ddots & \\ 0 & & \tan\theta_n\end{bmatrix}g$$

$$= \prod_{i=1}^{n} \frac{(\cos\theta_i)^{1/2}}{\sqrt{\cos\theta_i(e^{-i\theta_i})}}\, g$$

(Here $\operatorname{Re}\cos\theta_i e^{-i\theta_i} > 0$ and the square root is taken to be the standard branch.)

To prove 2) it is enough to prove it for $n = 1$. That is, if $\cos\theta \neq 0$, $\theta \in \mathbb{R}$ then

$$\mathcal{F}\mu(\tan\theta)\mathcal{F}^{-1}\alpha(\cos\theta)\mu(\tan\theta)g = \frac{|\cos\theta|^{1/2}}{\sqrt{\cos\theta(e^{-i\theta})}}\, g$$

for $g(x) = e^{-x^2/2}$.

Now

$$(\alpha(\cos\theta)\mu(\tan\theta)g)(x) = |\cos\theta|^{1/2} e^{i\tan\theta(\cos\theta)^2x^2} e^{-\cos^2\theta\, x^2/2}$$

$$= |\cos\theta|^{1/2} e^{-(\cos\theta(e^{-i\theta})x^2/2}$$

$$= |\cos\theta|^{1/2} h(x) \quad .$$

Note that $\mathrm{Re}(\cos\theta\, e^{-i\theta}) = \cos^2\theta > 0$. Hence

$$\hat{h}(x) = \frac{1}{\sqrt{\cos\theta(e^{-i\theta})}} e^{-x^2/2\cos\theta(e^{-i\theta})}$$

$$= \frac{1}{\sqrt{(\cos\theta)e^{-i\theta}}} e^{-x^2/2} e^{-i\tan\theta\, x^2/2}$$

Thus

$$\mu(\tan\theta)\mathcal{F}^{-1}\alpha(\cos\theta)\mu(\tan\theta)g = \frac{|\cos\theta|^{1/2}}{\sqrt{\cos\theta(e^{-i\theta})}} g \quad .$$

Hence

$$\mathcal{F}\mu(\tan\theta)\mathcal{F}^{-1}(\cos\theta)\mu(\tan\theta)g = \frac{|\cos\theta|^{1/2}}{\sqrt{\cos\theta(e^{-i\theta})}} g \quad .$$

This proves 2).

Now let $\tilde{h}$ be the Lie algebra of $\tilde{T}$. Then $\nu_*: \tilde{h} \to h$ (h the Lie algebra of T) is an isomorphism. Let

$$h_i = \begin{bmatrix} 0 & \mathrm{diag}(\delta_{1i},\ldots,\delta_{ni}) \\ -\mathrm{diag}(\delta_{1i},\ldots,\delta_{ni}) & 0 \end{bmatrix}$$

Let $\tilde{h}_i \in \tilde{h}$ be such that $\nu_*(\tilde{h}_i) = h_i$. Now $\tilde{T}$ has a neighborhood of I of the form

$$N_\epsilon = \{F\mu(\mathrm{diag}(\tan\theta_1,\ldots,\tan\theta_n))F^{-1}\alpha(\mathrm{diag}(\cos\theta_1,\ldots,\cos\theta_n))$$

$$\cdot\mu(\mathrm{diag}(\tan\theta_1,\ldots,\tan\theta_n)) \mid -\epsilon < \theta_i < \epsilon\}$$

$0 < \epsilon < \pi/2$, and $\nu: N_\epsilon \to T$ is injective. But

$$\exp(\nu_*(H)) = \nu(\exp H) .$$

Hence, if $-\epsilon < \theta_i < \epsilon$ then

$$\exp\left(\sum \theta_i \tilde{h}_i\right) = F\mu(\mathrm{diag}(\tan\theta_1,\ldots,\tan\theta_n))F^{-1}$$

$$\cdot\alpha(\mathrm{diag}(\cos\theta_1,\ldots,\cos\theta_n))\ (\mathrm{diag}(\tan\theta_1,\ldots,\tan\theta_n))$$

But this implies

$$\text{(3)} \qquad \exp\left(\sum\theta_i\tilde{h}_i\right)\cdot g = e^{i\,\Sigma_{j=1}^n \theta_j/2}\, g \quad .$$

$\theta_1,\ldots,\theta_n \in \mathbb{R}$.

Now $\nu(\exp(\Sigma\,\theta_i\tilde{h}_i)) = I$ if and only if $\exp(\Sigma\,\theta_i h_i) = I$. But this implies $\theta_i = 2\pi k_i$, $k_i \in \mathbb{Z}$. Now $\nu(\exp(\Sigma\,\theta_i\tilde{h}_i)) = I$ implies $\exp(\Sigma\,\theta_i\tilde{h}_i) = \lambda I$, $|\lambda| = 1$. But then, by 3)

$$\begin{aligned}\lambda &= e^{i\,\Sigma_{j=1}^{n}\,\theta_i/2} \\ &= e^{i\,\Sigma_{j=1}^{n}\,2\pi k_i/2} \\ &= \pm 1 \;.\end{aligned}$$

Taking, say $\theta_1 = 2\pi$, $\theta_i = 0$, $i > 1$, $\exp(2\pi\tilde{h}_i) = -I$. Thus $\operatorname{Ker}\nu = \{\pm I\}$. Q.E.D.

4. THE METAPLECTIC REPRESENTATION

In this section we look at $Mp(n,\mathbb{R})$ as an abstract Lie group. Then $\nu\colon Mp(n,\mathbb{R}) \to Sp(n,\mathbb{R})$ is a double covering.

Definition 4.1. The metaplectic representation of $Mp(n,\mathbb{R})$ is the unitary representation $(W, L^2(\mathbb{R}^n))$ where $W\colon Mp(n,\mathbb{R}) \to U(L^2(\mathbb{R}^n))$ is the realization of Definition 1.4.

We now give an abstract characterization of $(W, L^2(\mathbb{R}^n))$.

Theorem 4.2. Suppose that G is a connected Lie group and that $\eta\colon G \to Sp(n,\mathbb{R})$ is a non-trivial homomorphism. Suppose furthermore, that (π,H) is a non-trivial unitary representation of G and that there is an irreducible unitary representation of H_n, T, on H so that

$$\pi(g)T(z,t)\pi(g)^{-1} = T(\eta(g)\cdot z,t)$$
$$\text{for } g \in G, \quad z \in \mathbb{R}^{2n}, \quad t \in \mathbb{R} \ .$$

Then

1) There is $\gamma: G \to Mp(n,\mathbb{R})$ a Lie homomorphism, so that $\eta = \nu\circ\gamma$

2) (π,H) is equivalent with either $W\circ\gamma$ or $W^*\circ\gamma$ (here $W^*(g) = \sigma\circ W(g)\circ\sigma$ where $\sigma f(z) = \overline{f(z)}$).

<u>Proof</u>. We first note $\dim H = \infty$. Indeed, if $\dim H < \infty$ then Theorem 4.1 implies that $\dim H = 1$. But G is locally isomorphic with $Sp(n,\mathbb{R})$. Hence G has no non-trivial one-dimensional representations. (Cf. Wallach, [26], 8.2.)

But this implies that there is $A: H \to L^2(\mathbb{R}^n)$ so that

$$\left(A\circ T\left(\begin{bmatrix}x\\y\end{bmatrix},t\right)\circ A^{-1}f\right)(z) = \left(S_\lambda\left(\begin{bmatrix}x\\y\end{bmatrix},t\right)f\right)(z)$$
$$= e^{i\lambda(t+\langle x,z-\frac{1}{2}y\rangle)}f(z-y)$$

and $\lambda \neq 0$.

a) $\lambda > 0$.

Let

$$(V_\lambda f)(z) = \lambda^{-n/4} f(\lambda^{-1/2} z) .$$

Then

$$V_\lambda \circ S_\lambda(\lambda^{-1/2} z, \lambda^{-1} t) V_\lambda^{-1} = S_1(z,t) .$$

Set

$$A = V_\lambda \circ B .$$

Then if $U(g) = A \circ \pi(g) \circ A^{-1}$, then

$$U(g) \circ S_1(z,t) \circ U(g)^{-1} = S_1(\mu(g)z,t) .$$

But then $U: G \to W(Mp(n,\mathbb{R}))$, and since G is locally isomorphic with $Sp(n,\mathbb{R})$ we see that $U(g) = W(\gamma(g))$ with $\gamma: G \to Mp(n,\mathbb{R})$ a covering map.

b) $\lambda < 0$.

Again we may assume $H = L^2(\mathbb{R}^n)$, $T = S_\lambda$. By using $V_{|\lambda|}$ we may assume $T = S_{-1}$. Finally, $\sigma \circ S_{-1}(\lambda,t) \circ \sigma = S_1(x,t)$. Thus $\sigma \circ U \circ \sigma$ is equal to $W(\gamma(g))$. Thus U is equal to $W^*(\gamma(g))$. Q.E.D.

5. THE DIFFERENTIAL OF THE METAPLECTIC REPRESENTATION

Let V be a finite dimensional vector space over $\mathbb{R}$ and let $\langle\ ,\ \rangle$ be an inner product on V. As usual we set $\|v\|^2 = \langle v,v\rangle$. Let $(H,\langle\ ,\ \rangle)$ be a Hilbert space. If $x \in V$ and $f\colon V \to H$ is a function, then f is said to be differentiable at x if there is a linear map $\lambda\colon V \to H$ so that

$$f(x+v) - f(x) = \lambda(v) + h(v)\|v\|$$

with

$$\lim_{v\to 0} h(v) = 0 .$$

We recall (cf. Lang [17]) that if λ exists, it is unique and we use the notation

$$\lambda = Df(x)$$

(i.e., $\lambda(v) = Df(x)v$). In this context the chain rule says that if W is another finite dimensional vector space, if $g\colon W \to V$ is differentiable at x, and if f is differentiable at $g(x)$, then

$$D(f\circ g)(x) = Df(g(x))\circ Dg(x)$$

(cf. Lang [17]).

We identify the Lie algebra of $Mp(n,\mathbb{R})$ with the Lie algebra of $Sp(n,\mathbb{R})$, $\mathfrak{sp}(n,\mathbb{R})$.

<u>Lemma 5.1</u>. Let $\mathcal{P}_n = \{X \varepsilon Mn(\mathbb{R}) | \; {}^tX = X\}$. If $\varphi \varepsilon \mathcal{S}(\mathbb{R}^n)$ then the map $f\colon \mathcal{P}_n \to L^2(\mathbb{R}^n)$ given by

$$f(X) = W\left[\begin{bmatrix} I & X \\ 0 & I \end{bmatrix}\right]\varphi$$

is differentiable and

$$(Df(X)(v))(z) = i\langle vx, x\rangle/2 \; f(X)(z) \quad .$$

<u>Proof</u>. We use the following simple estimate:

1) $|e^{it} - 1 - it| \leq 3t^2$

for $t \varepsilon \mathbb{R}$.

In fact, if $|t| \leq 1$ then

$$\begin{aligned} |e^{it} - 1 - it| &\leq \sum_{k=2}^{\infty} \frac{|t|^k}{k!} \\ &\leq t^2 \sum_{k=2}^{\infty} \frac{1}{k!} \\ &= t^2(e-2) \\ &< 3t^2 \quad . \end{aligned}$$

If $|t| \geq 1$ then

$$\frac{|e^{it} - 1 - it|}{t^2} \leq \frac{2}{t^2} + \frac{|t|}{t^2} \leq 3 .$$

This clearly implies 1).

We now prove the lemma. We first note that

$$f(x+v) = W\begin{bmatrix} I & x \\ 0 & I \end{bmatrix} f(v) .$$

This says that

$$(*) \qquad \left\| \frac{f(x+v) - f(x) - i\langle v\cdot,\cdot\rangle/2\, f(x)}{\|v\|} \right\|^2$$

$$= \frac{1}{\|v\|^2} \int_{\mathbb{R}^n} |e^{i\langle vz,z\rangle/2} - 1 - i\langle vz,z\rangle/2|^2\, |\varphi(x)|^2\, dz$$

$$\leq \frac{3}{\|v\|^2} \int_{\mathbb{R}^n} |\langle vz,z\rangle|^4/4\, |\varphi(z)|^2\, dz .$$

Now

$$|\langle vz,z\rangle| \leq \|vz\|\,\|z\| \leq \|z\|^2\,\|v\|$$

Hence

$$(*) \leq \frac{3}{4}\|v\|^2 \int_{\mathbb{R}^n} \|z\|^4\, |\varphi(z)|^2\, dz .$$

This proves the lemma.

Corollary 5.2. Let $\varphi \in \mathcal{S}(\mathbb{R}^n)$ and let $f\colon \mathcal{P}_n \to L^2(\mathbb{R}^n)$ be given by

$$f(X) = W\begin{bmatrix} I & 0 \\ X & I \end{bmatrix}$$

Then f is differentiable at every $X \in \mathcal{P}_n$ and

$$Df(x)\cdot v = W\begin{bmatrix} I & 0 \\ X & I \end{bmatrix} \cdot \left(\frac{i}{2} \sum_{ij} v_{ij} \frac{\partial^2}{\partial z_i \partial z_j} \varphi \right) ,$$

$$V = (v_{ij}) .$$

Proof.

$$W\begin{bmatrix} I & 0 \\ -X & I \end{bmatrix} = \mathcal{F} W \left(\begin{bmatrix} I & X \\ 0 & I \end{bmatrix} \right) \mathcal{F}^{-1} .$$

Now use Lemma 5.1.

Lemma 5.3. Let, for $\varphi \in \mathcal{S}(\mathbb{R}^n)$, $f\colon Mn(\mathbb{R}) \to L^2(\mathbb{R}^n)$ be given by

$$f(x) = W\begin{bmatrix} \exp X & 0 \\ 0 & \exp(-{}^tX) \end{bmatrix} \varphi$$

Then f is differentiable at every $X \in Mn(\mathbb{R})$ and

$$Df(x)\cdot v = \frac{1}{2}\,tr(V)f(X) + W\begin{bmatrix}\exp X & 0\\ 0 & \exp(-{}^tX)\end{bmatrix}\cdot\left(\sum_{ij} z_j v_{ji}\frac{\partial\varphi}{\partial z_i}\right)$$

$$v = (v_{ij})$$

This is an easy consequence of Taylor's theorem with integral remainder and the change of variables formula. We leave it as an exercise to the reader.

Recall that $sp(n,\mathbb{R}) = \bar{n} \oplus m \oplus n$ see 1) in the proof of Theorem 3.2. If $Z \in sp(n,\mathbb{R})$ then $Z = Z_{\bar{n}} + Z_m + Z_n$ relative to this decomposition with

$$Z_{\bar{n}} = \begin{bmatrix} I & 0\\ X(Z) & I\end{bmatrix}, \quad Z_m = \begin{bmatrix} m(Z) & 0\\ 0 & -{}^tm(Z)\end{bmatrix}, \quad Z_n = \begin{bmatrix} I & Y(Z)\\ 0 & I\end{bmatrix}$$

<u>Theorem 5.4</u>. If $\varphi \in S(\mathbb{R}^n)$ then the map $f\colon sp(n,\mathbb{R}) \to L^2(\mathbb{R}^n)$ given by $f(Z) = W(\exp Z)\varphi$, $(\exp\colon sp(n,\mathbb{R}) \to Mp(n,\mathbb{R})$ is differentiable at all Z and

$$Df(Z)(v) = W(\exp Z)\cdot\left(\frac{i}{2}\sum_{k,\ell} X(Z)_{k\ell}\frac{\partial^2\varphi}{\partial z_k\partial z_\ell} + \frac{1}{2}\,tr(m(Z))\varphi\right.$$

$$\left. + \sum_{k,\ell} z_k m(Z)_{k\ell}\frac{\partial\varphi}{\partial z_\ell} + \frac{i}{2}\langle Y(Z)\cdot,\cdot\rangle\,\varphi\right).$$

Proof.

$$\exp Z = \bar{n}(\exp Z)\cdot m(\exp Z)n(\exp Z)$$

(see 1) in the proof of Lemma 3.1) for Z sufficiently small (here we identify $\bar{N}MN$ with the connected component of the inverse image of $\bar{N}MN$ in $Mp(n,\mathbb{R})$ containing the identity). Now use the chain rule and 5.1, 5.2, 5.3.

6. THE HERMITE FUNCTIONS

In this section we show how the metaplectic representation of $Mp(1,\mathbb{R})$ can be used to give a development of the theory of Hermite functions in terms of Fourier series.

We note that

$$Sp(1,\mathbb{R}) = SL(2,\mathbb{R}) \quad .$$

We identify the Lie algebra of $Mp(1,\mathbb{R})$ with that of

$$\mathfrak{sl}(2,\mathbb{R}) = \left\{ \begin{bmatrix} a & b \\ c & d \end{bmatrix} \middle| \ a+d = 0,\ a,b,c,d \in \mathbb{R} \right\} \quad .$$

We look at the element

$$J = \begin{bmatrix} 0 & 1 \\ -1 & 0 \end{bmatrix} \qquad \text{in } \mathfrak{sl}(2,\mathbb{R}) \quad .$$

Let us denote by $\exp$ the exponential map $\mathfrak{sl}(2,\mathbb{R}) \to SL(2,\mathbb{R})$ given by

$$\exp X = e^X = \sum_{k=0}^{\infty} \frac{1}{k!} X^k$$

and by Exp the exponential map of $\mathfrak{sl}(2,\mathbb{R})$ to $Mp(1,\mathbb{R})$. Let

$$T = \exp(\mathbb{R}J)$$

and

$$\tilde{T} = \operatorname{Exp}(\mathbb{R}J) \quad .$$

Then

$$\operatorname{Ker}(\exp\colon \mathbb{R}J \to T) = 2\pi\mathbb{Z}J$$

and

$$\operatorname{Ker}(\operatorname{Exp}\colon \mathbb{R}J \to \tilde{T}) = 4\pi\mathbb{Z}J \quad .$$

If $\varphi \in L^2(\mathbb{R})$ then $\tilde{T} \to L^2(\mathbb{R})$ given by $z \mapsto W(z)\varphi$ is a continuous function. Theorem 5.4 implies that if $\varphi \in \mathcal{S}(\mathbb{R})$ then $z \mapsto W(z)\varphi$ is a C^∞ function from $\tilde{T}$ to $L^2(\mathbb{R})$.

Let

$$\xi_k(\operatorname{Exp} tJ) = e^{i\,kt/2}$$

$$\text{for } k \in \mathbb{Z} \quad ,$$

Then ξ_k is well defined and defines a character of $\tilde{T}$. It is easily seen that if ξ is a character of $\tilde{T}$, then $\xi = \xi_k$ for some $k \in \mathbb{Z}$. By the above we see that if $\varphi \in \mathcal{S}(\mathbb{R})$, then

$$W(z)\varphi = \sum \xi_k(z)\varphi_k$$

with $\varphi_k \in L^2(\mathbb{R})$,

$$\varphi_k = \frac{1}{4\pi} \int_0^{4\pi} \overline{\xi_k(\mathrm{Exp}(-tJ))}\, W(z)\varphi \, dz$$

and the convergence is absolute.

Set

$$P_k\varphi = \varphi_k$$

for $\varphi \in S(\mathbb{R})$.

Then noting that

$$W(z)\varphi_k = \xi_k(z)\varphi_k \quad ,$$

$$k \in \mathbb{Z}, \quad z \in \tilde{T} \quad ,$$

it is clear that $\langle\varphi_k, \varphi_\ell\rangle = 0$ if $k \neq \ell$. Hence $\|\varphi_k\| \leq \|\varphi\|$.

This implies that P_k extends to a bounded operator from $L^2(\mathbb{R})$ to $L^2(\mathbb{R})$. Clearly, $P_k^2 = P_k$.

Now Theorem 5.4 implies that if $\varphi \in S(\mathbb{R})$ then

$$\frac{d}{dt} W(\mathrm{Exp}\ tJ)\varphi\Big|_{t=0} = -\frac{i}{2}\frac{d^2}{dx^2}\varphi + \frac{i}{2}x^2\varphi$$

$$= \frac{i}{2}\left(-\frac{d^2}{dx^2} + x^2\right)\varphi \quad .$$

<u>Definition 6.1</u>. The differential operator

$$H = -\frac{d^2}{dx^2} + x^2$$

is called the <u>Hermite</u> operator.

Suppose that $f \varepsilon P_k(L^2(\mathbb{R}))$. Then

$$W(\text{Exp } tJ)f = e^{ikt/2} f \quad .$$

If $\varphi \varepsilon S(\mathbb{R})$ then

a) $$\frac{d}{dt}_{t=0} <W(\text{Exp } tJ)f, \ > = \frac{ik}{2} <f,\varphi>$$

b) $$\frac{d}{dt}_{t=0} <W(\text{Exp } tJ)f,\varphi> = \frac{d}{dt}_{t=0} <f,W(\text{Exp}(-tJ)) \ >$$

$$= -\frac{i}{2} <f,H\varphi>$$

Combining a) and b) we find

c) If $f \varepsilon P_k L^2(\mathbb{R})$ and $\varphi \varepsilon S(\mathbb{R})$ then

$$<f,H\varphi> = -k<f,\varphi> \quad .$$

c) says that f is a <u>weak solution</u> of the equation $Hf = -kf$. The general theory of elliptic operators implies that f is a solution of $Hf = -kf$. That is, f is of class C^∞ and

$$-\left(\frac{d^2}{dx^2} f\right) + x^2 f = -kf \quad .$$

However, we will give a proof of this fact in the Appendix to this chapter (see Theorem A.5.1). We now know that $\dim P_k L^2(\mathbb{R}) \leq 2$.

In the course of the proof of Theorem 3.2 we showed that if $g(x) = \exp(-x^2/2)$, $W(\text{Exp } tJ)g = e^{it/2} g$. Hence,

$$\frac{i}{2} Hg = \frac{i}{2} g \quad .$$

That is,

$$Hg = g$$

(Of course, this could be proved by the obvious computation.) We are interested in seeing if there is a second solution to this equation in $L^2(\mathbb{R})$. Since g is nowhere zero, the usual techniques for computing second solutions is to look at $f = ug$. Then

$$Hf - f = -\left(\frac{d^2u}{dx^2} g + 2 \frac{du}{dx} \frac{dg}{dx}\right) \quad .$$

$$\frac{dg}{dx} = -xg \quad .$$

Hence, we find

$$\frac{d^2u}{dx^2} = 2x\frac{du}{dx} \quad .$$

We therefore see that

$$\frac{du}{dx} = ce^{x^2} \quad .$$

Hence, if $Hf = f$,

$$f = Ag + Bug$$

with

$$\frac{du}{dx} = e^{x^2} \quad , \qquad u(0) = 1 \quad .$$

It is easy to see that $u(x) > e^{x^2/2}$. Thus if $f \in L^2(\mathbb{R})$, $B = 0$. We have therefore shown that

$$P_1 L^2(\mathbb{R}) = \mathbb{C}g \quad .$$

To find other eigenfunctions for H we use the definition of the metaplectic representation.

We leave it to the reader to check that if $\varphi \in \mathcal{S}(\mathbb{R})$, then

I) $$\frac{d}{dt} S_1\left(t\begin{bmatrix}a\\b\end{bmatrix},0\right)\varphi\Big|_{t=0} = aix\varphi - b\frac{d\varphi}{dx} \quad .$$

Also,

$$W(\mathrm{Exp}(\theta J))S_1\left(t\begin{bmatrix}a\\b\end{bmatrix},0\right)W(\mathrm{Exp}(-\theta J)) = S_1\left(t\begin{bmatrix}\cos\theta a + \sin\theta b\\ -\sin\theta a + \cos\theta b\end{bmatrix},0\right)$$

Hence taking derivatives we find that if $\varphi \in S(\mathbb{R})$, then

$$W(\mathrm{Exp}(\ J))\left(-b\,\frac{d}{dx} + aix\right)W(\mathrm{Exp}(-\ J))\varphi$$

$$= \quad ((\sin\theta a - \cos\theta b)\,\frac{d}{dx} + i((\cos\theta)a + (\sin\theta)b)x)\varphi$$

In particular, if $a = i$, $b = 1$, then

$$W(\mathrm{Exp}(\theta J))\ -\frac{d}{dx} - x\ W(\mathrm{Exp}(-\theta J))\varphi \quad = \quad -\left(e^{-i\theta}\,\frac{d}{dx} - e^{-i\theta}\ x\right)\varphi\,.$$

Hence if

$$A^{-} \quad = \quad \frac{d}{dx} + x$$

then

$$W(\mathrm{Exp}(\theta J))A^{-}\ W(\mathrm{Exp}(-\theta J))\varphi \quad = \quad e^{-i\theta}\ A^{-}\varphi$$

for $\varphi \in S(\mathbb{R})$.

Using $a = -i$ and $b = 1$ then

$$W(\mathrm{Exp}(\theta J))A^{+}\ W(\mathrm{Exp}(-\theta J))\varphi \quad = \quad e^{i\theta}\ A^{+}\varphi\ .$$

II) $\quad H \quad = \quad -\frac{1}{2}\,(A^{+}A^{-} + A^{-}A^{+})$

II) is a simple computation. Also, if $\psi,\ \ \in S(\mathbb{R})$, then

$$<A^{+}\varphi,\psi> \quad = \quad -<\varphi,A^{-}\psi>$$

The above computations imply

III) $\quad (H,A^{+}] \quad = \quad 2A^{+} \quad$ and $\quad (H,A^{-}] \quad = \quad -2A^{-}$.

III) implies that $(A^+)^k g \in P_{2k+1}(L^2(\mathbb{R}))$. We note that $A^- g = 0$.

IV) If $P_k L^2(\mathbb{R}) \neq 0$, then $k > 0$.

This follows from Lemma A.5.2. Also, Lemma A.5.2 implies immediately that

V) If $f \in P_k(L^2(\mathbb{R}))$ then $A^- f \in P_{k-2}(L^2(\mathbb{R}))$.

We note that if $k \neq 1$ and $f \in P_k(L^2(\mathbb{R}))$, $f \neq 0$ then $A^- f \neq 0$. Indeed, $\mathbb{C}g = \{h |\ A^- h = 0\}$. This implies that if $f \in P_{2k}(L^2(\mathbb{R}))$ then

$$(A^-)^k f \in P_0(L^2(\mathbb{R})) = (0)$$

(by IV)), hence $f = 0$. Thus

VI) $P_{2k}(L^2(\mathbb{R})) = 0$.

If $f \in P_{2k+1}(L^2(\mathbb{R}))$, $k \geq 0$, then

$$(A^-)^h f \in P_1(L^2(\mathbb{R})) = \mathbb{C}g \quad .$$

Since $A^-)^k|_{P_{2k+1}(L^2(\mathbb{R}))}$ is injective we see that III) implies

V) $P_{2k}(L^2(\mathbb{R})) = P_j(L^2(\mathbb{R})) = 0$

for $k \in \mathbb{Z}$, $j \leq 0$.

$$P_{2k+1}(L^2(\mathbb{R})) = \mathbb{C}(A^+)^k g \quad , \qquad k = 0,1,\ldots$$

Definition 6.2. Set

$$\tilde{h}_k = (A^+)^k g \quad .$$

Then $\tilde{h}_k$ is called the k-th *unnormalized Hermite function*.
Set

$$h_k = \frac{1}{||h_k||} \tilde{h}_k \quad ,$$

$k = 0,1,\ldots$ Then h_k is called the k-th *Hermite function*.

We have proved

Lemma 6.3. The functions $\{h_k\}$ form an orthonormal basis of $L^2(\mathbb{R})$.

We also note for future reference

Lemma 6.4. If $\varphi \varepsilon S(\mathbb{R})$ then $|<\varphi,h_\ell>| \leq (2\ell+1)^{-k}||H^k\varphi||$, $k = 0,1,\ldots$

Proof.

$$\begin{aligned} <H^k\varphi,h_\ell> &= <\varphi,H^k h_\ell> \\ &= (2\ \ +1)^k<\varphi,h_\ell> \end{aligned}$$

Thus,

$$\begin{aligned} (2\ \ +1)^k|<\varphi,h_\ell>| &\leq ||h_\ell||\,||H^k\varphi|| \\ &= ||H^k\varphi|| \end{aligned}$$

APPENDIX TO CHAPTER 5

The purpose of this appendix is to give an elementary proof of the following theorem and Theorem A.5.3.

Theorem A.5.1. Let $q\colon \mathbb{R} \to \mathbb{R}$ be a C^∞ function. Set

$$Tf = -\frac{d^2f}{dx^2} + qf$$

$$\text{for } f \in C^2(\mathbb{R}) \quad .$$

If $f \in L^2(\mathbb{R})$ and $\langle f, Tg\rangle = 0$ for all $g \in C_c^\infty(\mathbb{R})$, then $f \in C^\infty(\mathbb{R})$ and $Tf = 0$.

Proof. Let $-\infty < a < b < \infty$, $a,b \in \mathbb{R}$. It is enough to show that if $f \in L^2([a,b])$ satisfies

$$\int_a^b f(t)\,\overline{Tg(t)}\,dt = 0$$

$$\text{for all } g \in C_c^\infty((a,b)) \quad ,$$

then there exists $\psi \in C^\infty(a,b)$ so that $f = \psi$, a.e. on (a,b) and $T\psi = 0$.

Let

$$V_0 = \{\varphi \in C^\infty([a,b]) \mid T\varphi = 0\}$$

Then $V_0 \subset L^2([a,b])$ and since $\dim V_0 = 2$, V_0 is closed in $L^2([a,b])$.

a) If $\varphi \in C_c^\infty((a,b))$ and $\varphi \perp V_0$, then there exists $g \in C_c^\infty((a,b))$ so that $Tg = \varphi$.

To prove a) we note that the existence and uniqueness theorem for ordinary differential equations implies that there exists a unique $g \in C^\infty([a,b])$ so that $Tg = \varphi$, $g(a) = g'(a) = 0$. If we can show that $g(b) = g'(b) = 0$, then the existence and uniqueness theorem will imply that if $\varphi(x) = 0$ on $[a, a+\epsilon)$ and $(b-\epsilon, b]$ then $g(x) = 0$ on $[a, a+\epsilon)$ and $(b-\epsilon, b]$ which would prove a). If $\psi \in V_0$, then

$$
\begin{aligned}
0 &= \int_a^b g(t)\,\overline{T\psi(t)}\,dt \\
&= \int_a^b g(t)(\overline{-\psi''(t)} + q(t)\overline{\psi(t)})\,dt \\
&= -\int_a^b g(t)\overline{\psi''(t)}\,dt + \int_a^b q(t)g(t)\overline{\psi}(t)\,dt \\
&= -\int_a^b \frac{d}{dt}(g(t)\overline{\psi}'(t))\,dt + \int_a^b g'(t)\overline{\psi'(t)}\,dt \\
&\quad + \int_a^b q(t)g(t)\overline{\psi}(t)\,dt
\end{aligned}
$$

$$= -g\overline{\psi}' + (g'\overline{\psi})' \Big|_a^b + \int_a^b (Tg)\overline{\psi}\, dt$$

$$= -g(b)\overline{\psi}'(b) + g'(b)\overline{\psi}(b) + \langle\varphi,\psi\rangle \quad .$$

But $\langle\varphi,\psi\rangle = 0$ by hypothesis. Hence,

$$-g(b)\overline{\psi}'(b) + g'(b)\overline{\psi}(b) = 0 \quad .$$

Now, $\psi \in V_0$ is arbitrary. Pick ψ so that $\psi(b) = g'(b)$, $\psi'(b) = -g(b)$. Then $|g(b)|^2 + |g'(b)|^2 = 0$. a) now follows.

b) Let f be as in the statement of the theorem. If $\varphi \in C_c^\infty(a,b)$ and $\varphi \perp V_0$, then $\varphi \perp f$.

Indeed, $\varphi = Tg$ with $g \in C_c^\infty(a,b)$ by a). Hence $\langle\varphi,f\rangle = \langle Tg,f\rangle = 0$.

Let ψ_1,ψ_2 be an orthonormal basis of V_0. Let $\eta_1,\eta_2 \in C_c^\infty(a,b)$ be such that $\|\eta_i-\psi_i\| < \epsilon$, $i = 1,2$ and let ϵ be so small that $\det(\langle\eta_i,\psi_j\rangle) \neq 0$. Let

$$(b_{ij}) = (\langle\eta_i,\psi_j\rangle)^{-1} \quad .$$

Suppose that $h \in V_0^\perp$. Let $h_j \to h$, a.e. $h_j \in C_c^\infty((a,b))$, $\|h_j\| < C$. Set

$$g_j = h_j - \sum_{k,\ell} b_{k,\ell}\eta_\ell\langle h_j,\psi_k\rangle \quad .$$

Then $<g_j,\psi_i> = 0$, $i = 1,2$. Hence, $g_j \in V_0^{\perp}$. b) implies that $<g_j,f> = 0$.

$$\lim_{j\to\infty} <h_j,\psi_k> = <h,\psi_k>$$
$$= 0 \quad .$$

Hence

$$0 = \lim_{j\to\infty} <g_j,f>$$
$$= \lim_{j\to\infty} <h_j,f>$$
$$= <h,f>$$

This implies that $f \in (V_0^{\perp})^{\perp} = V_0$. Q.E.D.

Lemma A.5.2. Let $f \in L^2(\mathbb{R}) \cap C^{\infty}(\mathbb{R})$, $f \neq 0$ be a solution to

$$-\frac{d^2f}{dx^2} + x^2f = kf$$

with $k \in \mathbb{R}$. Then $k > 0$ and xf and f' are in $L^2(\mathbb{R})$.

Proof. $d^2f/dx^2 = (x^2-k)f$. Since $k \in \mathbb{R}$, $\mathrm{Re}f$ and $\mathrm{Im}f$ satisfy the hypothesis of the lemma. Hence we may assume that f is real valued.

Let $x_0^2 - k > 1$. Suppose that $x_0 \leq a < b$ and $f(a) = f(b) = 0$ but $f(x) \neq 0$, $a < x < b$. We may assume $f(x) > 0$, $a < x < b$. This implies $f''(x) > 0$ for $a < x < b$. Now there is $a < c < b$ so that $f'(c) = 0$. Hence $f'(x) < 0$ for $a < x < c$. But then $f(x) < 0$, $a < x < c$. This contradiction implies f has at most one zero for $x \geq x_0$. We may thus assume $f(x) > 0$ for $x \geq x_0$. If $f'(c) = 0$ for $c > x_0$, then $f(x) < 0$ for $x_0 < x < c$. Hence $f'(x) \neq 0$ for all $x > x_0$. If $f'(x) > 0$ for $x > x_0$, then f is increasing for $x \geq x_0$. This implies $f(x) \geq f(x_0) = c_0 > 0$. Hence $f \notin L^2(\mathbb{R})$. Thus $f'(x) < 0$, $x > x_0$.

We have shown

a) There is x_0 so that if $x > x_0$, then $f'(x)f(x) < 0$.

A similar argument shows that

b) there is x_1 so that if $x < x_1$, then $f'(x)f(x) > 0$.

Let $a < x_1$, $b > x_0$. Then

$$\int_a^b \left(-\frac{d^2f}{dx^2} + x^2 f\right) f = k \int_a^b f^2 .$$

This implies

c) $$-f'f \Big|_a^b + \int_a^b (f')^2 + \int_a^b (xf)^2 = k \int_a^b f^2 .$$

c) clearly implies that $k > 0$ and

$$\int_a^b (f')^2 < k \int_a^b f^2 \quad , \quad \int_a^b (xf)^2 < k \int_a^b f^2$$

Taking the limit $a \to -\infty$, $b \to \infty$ gives the result.

Theorem A.5.3. Let h_n be the nth Hermite function. If $\varphi \in S(\mathbb{R})$, then

$$\sum_{n=0}^{\infty} \langle \varphi, h_n \rangle \, h_n$$

converges to φ in the sense of $S(\mathbb{R})$.

To prove this theorem we need several simple lemmas.

Lemma A.5.4. Let $\|f\|$ denote the L^2 -norm of $f \in L^2(\mathbb{R})$. If $p,q \in \mathbb{Z}^+$, then there exists a constant $C_{p,q}$ so that if $u \in S(\mathbb{R})$,

$$|x^p u^{(q)}(x)| \le C_{p,q}(\|x^p u^{(q)}\| + \|x^{p+1} u^{(q+1)}\|)$$

Proof.

$$x^{p+1} u^{(q)}(x) = p + 1 \int_0^x x^p u^{(q)}(t)\, dt + \int_0^x t^{p+1} u^{(q+1)}(t)\, dt$$

Hence

$$|x^{p+1}u^{(q)}(x)| \leq (p+1)(x)\|x^p u^{(q)}\| + |x|\,\|x^{p+1}u^{(q+1)}\|$$

by Schwarz inequality. Dividing through by x gives the estimate. ($C_{p+q} = p+1$ will do.)

Lemma A.5.5. Let

$$Hf = -\frac{d^2f}{dx^2} + x^2 f .$$

If $k \in \mathbb{Z}$, $k \geq 0$, then there is a constant C_k so that if $u \in \mathcal{S}(\mathbb{R})$, then

$$\|x^p u^{(q)}\|^2 \leq C_{p+q} \sum_{j=0}^{p+q} \langle H^j u, u\rangle .$$

Proof.

a) $$H^k = \sum_{j=0}^{k} \binom{k}{j}(-1)^j x^{2k-2j} \frac{d^{2j}}{dx^{2j}} + \sum_{p+q<2k} C_{p,q,k} x^p \frac{d^q}{dx^q}$$

a) can be proved by noting that H^k is a sum of terms of the form

$$A_1 \cdots A_k , \quad A_i = -\frac{d^2}{dx^2} \text{ or } x^2 .$$

If we wish to rearrange the A_i so that all of the x^2 terms come first and the $-d^2/dx^2$ come last, we must take commutators

$$\left[\frac{d^2}{dx^2}, x^2\right] = 2I + 4x\frac{d}{dx} .$$

We prove the lemma by induction on $k = p+q$. If $k = 0$ the result is clear. If $k = 1$ then

$$\langle Hu,u\rangle = \|u'\|^2 + \|xu\|^2 .$$

Hence $C_1 = 1$ will do.

Suppose we have proved the result for $0 \leq r < k$. Then

$$\langle H^k u,u\rangle = \sum_{j=0}^{k} \binom{k}{j}(-1)^j \langle x^{2k-2j}u^{(2j)}, u\rangle$$

$$+ \sum_{p+q<2k} C_{p,q,k}\langle x^p u^{(q)}, u\rangle$$

b) If $p+q < 2k$ then $\langle x^p u^{(q)}, u\rangle \leq D_{p,q} \sum_{j=0}^{k-1} \langle H^k u,u\rangle$.

Indeed, let

$$(r,s) + (\alpha,\beta) = (p,q) ,$$

$r,s,\alpha,\beta \geq 0$ in $\mathbb{Z}$ and $r+s < k$, $\alpha,\beta < k$. Then

$$\langle x^p u^{(q)},u\rangle = \left\langle x^r x^\alpha \frac{d^s}{dx^s} u^{(\beta)}, u\right\rangle$$

$$= \left\langle x^\alpha \frac{d^s}{dx^s} u^{(\beta)}, x^r u\right\rangle$$

$$= (-1)^s \sum_{j=0}^{s} \binom{s}{j}\binom{\alpha}{j} j! \left\langle x^{\alpha-j}u^{(\beta)}, \left[\frac{d^{s-j}}{dx^{s-j}}\right] x^r u\right\rangle$$

Hence

$$|\langle x^p u^{(q)}, u\rangle| \le \sum_{j=0}^{s} \binom{s}{j}\binom{\alpha}{j} j! \cdot \sum_{\gamma=0}^{s-j} \binom{r}{\gamma} \gamma! \, |\langle x^{\alpha-j} u^{(\beta)}, x^{r-\gamma} u^{(\gamma)}\rangle|$$

$$- \frac{1}{2} \sum_{j=0}^{s} \binom{s}{j}\binom{\alpha}{j} j! \sum_{\gamma=0}^{s-j} \binom{r}{\gamma} \gamma! \Big(\|x^{\alpha-j} u^{(\beta)}\|^2$$

$$+ \|x^{r-\gamma} u^{(\gamma)}\|^2 \Big)$$

Now apply the inductive hypothesis. We need to look at

$$\sum_{j=0}^{k} \binom{k}{j} (-1)^j \langle x^{2k-2j} u^{(2j)}, u\rangle$$

$$= \sum_{j=0}^{k} \binom{k}{j} \sum_{s=0}^{j} \binom{j}{s}\binom{2k-2j}{s} s! \, \langle x^{k-j} u^{(j)}, x^{k-j-s} u^{(j-s)}\rangle$$

If $s \ge 1$, $s \le 2k-2j$ then

$$|\langle x^{k-j} u^{(j)}, x^{k-j-s} u^{(j-s)}\rangle| = |\langle x^{k-j-1} u^{(j)}, x^{k-j-s+1} u^{(j-s)}\rangle|$$

$$\le \frac{1}{2} (|x^{k-j-1} u^{(j)}|^2$$

$$+ |x^{k-j-s+1} u^{j-s}|^2)$$

Now

$$k - j - s + 1 + j - s = k - 2s + 1$$

$$\le k - 1 \quad .$$

Hence

$$|<x^{k-j}u^{(j)}, x^{k-j-s}u^{(j-s)}>| \leq \max(C_{k-j}, C_{k-2s+1}) \sum_{j=0}^{k-1} <H^k u,u>$$

This implies that

$$\sum_{j=0}^{k} \binom{k}{j} \| x^{k-j}u^{(j)} \|^2 = <H^k u,u> + B_k(u)$$

$$|B_k(u)| \leq V_k \sum_{j=0}^{k-1} <H^j u,u>$$

Q.E.D.

<u>Proof of Theorem A.5.3</u>. We must show that if $p \geq 0$, $q \geq 0$ are given

$$\lim_{N\to\infty} \sup_x \left| x^p \left(\sum_{n=0}^{N} <\varphi, h_n> h_n \right)^{(q)} - x^p \varphi^{(q)} \right| = 0 \quad .$$

By Lemma A.5.4 it is enough to show that for all $p,q \geq 0$, $p,q \in \mathbb{Z}$,

$$\lim_{N\to\infty} \left\| x^p \left(\sum_{j=0}^{N} <\varphi, h_n> h_n \right)^{(q)} - x^p \varphi^{(q)} \right\| = 0 \quad .$$

But now it is enough to show that if $k \geq 0$, $k \in \mathbb{Z}$

$$\lim_{N\to\infty} \left\| H^k\left(\sum_{n=0}^{N} <\varphi,h_n>h_n\right) - H^k\varphi \right\| = 0$$

But

$$H^k h_n = (2n+1)^k h_n$$

and

$$<H^k\varphi,h_n> = (2n+1)^k<\varphi,h_n> \quad .$$

Hence, since

$$\lim_{N\to\infty} \sum_{n=0}^{N} <H^k\varphi,h_n>h_n = H^k\varphi$$

in $L^2(\mathbb{R})$. The result now follows. Q.E.D.

Chapter 6

QUANTIZATION

In this chapter we give a working definition of quantization. One can see (see Section 1) that the "definition" is quite vague. It contains as special cases the classical quantization of early quantum mechanics and the Kirillov association of unitary representations to orbits in the co-adjoint representation of a nilpotent Lie group. (Actually, the classical quantization is a special case of the Kirillov association.) We make no attempt to include the Auslander, Kostant theory [1] or the general Blattner, Kostant, Sternberg theory [3]. To do this would involve massive changes in the definition. (After all, it is just a working definition.) It is our feeling that the mathematical community has not as yet come up with a satisfactory definition of quantization. Thus, though our "definition" is clearly unsatisfactory, it at least has the advantage of being simple-minded and including the old quantum mechanical ideas of quantization and polarization. Chapter 7, The Kirillov Theory, can be thought of as an expansion of the "Kirillov quantization" discussed in Section 4 for nilpotent groups.

1. QUANTIZATION

In this section we give an attempt at an abstract working definition of "quantization". We will see that the definition will need modification.

Let (M,ω) be a symplectic manifold. We assume that ω is integral. We also assume $H^1(M,\mathbb{R}) = 0$. Let $A \subset C^\infty(M;\mathbb{R})$ be a space of functions closed under Poisson bracket. Let $B \subset A$ be a subalgebra of A. We assume that

1) If $p \in M$, then there exists U an open neighborhood of p in M and elements $f_1,\ldots,f_{2n} \in B$ so that

$$f_1\Big|_U, \ldots, f_{2n}\Big|_U$$

form a system of local coordinates on U.

Let (L,∇,h) be a triple of a line bundle L over M, a connection ∇ on L so that

$$\frac{1}{2\pi i}\,\mathrm{curv}(L,\nabla) = \omega \quad ,$$

and a Hermitian structure h on L.

Let δ be the corresponding prequantization of $C^\infty(M;\mathbb{R})$.

By a <u>quantization</u> of (A,B) relative to (L,∇) we mean a nonzero subspace of $C^\infty(M;L)$, H_∞, with a pre-Hilbert space structure $\langle\, ,\rangle$ and a map $\phi\colon A \to \mathrm{End}(H_\infty)$ so that

I) $\phi(f)$ is formally skew adjoint for $f \in A$. That is, $\langle\phi(f)\alpha,\beta\rangle = -\langle\alpha,\phi(f)\beta\rangle$.

II) $\phi(f_1)\phi(f_2) - \phi(f_2)\phi(f_1) = \phi(\{f_1,f_2\})$, $f_i \in A$, $i = 1,2$.

III) $\phi(f) = \delta(f)$ if $f \in B$.

IV) If H is the Hilbert space completion of $(H_\infty,\langle\ ,\ \rangle)$ and if $f \in B$ then there exists a unitary representation ρ_f of $\mathbb{R}$ on H so that if $\varphi \in H_\infty$ then

$$\lim_{t\to\infty} \frac{\rho_f(t)\varphi - \varphi}{t} = \phi(f)\varphi$$

V) If $A\colon H \to H$ is a bounded operator so that $A\circ\rho_f(t) = \rho_f(t)\circ A$ for all $t \in \mathbb{R}$, $f \in B$ then $A = cI$.

We will give the classical example in Section 6.2. We note that this definition does not coincide with the usual definitions in the mathematical literature (cf. Blattner [3]). It is also hopelessly abstract since it does not give any technique for finding H_∞ nor does it say how to put a pre-Hilbert space structure on H_∞. However, V) implies that the pre-Hilbert space structure is unique. We also note that this definition does not allow for the quantization of a single function. A worse drawback is that even though one might be interested in a finite dimensional Lie subalgebra of $C^\infty(M;\mathbb{R})$, $\mathcal{C}$, one must still find a $\mathcal{B}$ satisfying 1) above and an $\mathcal{A} \supset \mathcal{C}$, $\mathcal{A} \supset \mathcal{B}$. It is, however, broad enough for our purposes.

In practice the space H_∞ is given by a system of differential equations or functional equations. When H_∞ is given by differential equations, we are closer to the Blattner, Kostant, Sternberg (cf. Blattner [3]) (BKS) notion of quantization. The corresponding differential equations are usually <u>polarizations</u> in the (BKS) theory. The idea of using functional equations also to define H comes from Cartier [5], in which he realizes the basic representation of the Heisenberg group in a third way (it is given in the Appendix to Chapter 4 of these notes). Finally (as we shall see), in practice one has $\mathcal{B}$, H_∞ and $\langle\, ,\rangle$ and looks for the largest $\mathcal{A} \supset \mathcal{B}$ so that I),...,V) are satisfied.

2. THE CLASSICAL EXAMPLE

Let $M = \mathbb{R}^{2n}$ with linear coordinates $x_1,\dots,x_n$, $y_1,\dots,y_n$ (i.e., $x_i(z_1,\dots,z_{2n}) = z_i$, $i = 1,\dots,n$, $y_i{*}z_1,\dots,z_{2n}) = z_{n+i}$, $i = 1,\dots,n$). Let

$$\omega = \sum dx_i \wedge dy_i \quad .$$

Let $\mathcal{B}$ be the space of all polynomials of degree at most one. Then $\mathcal{B}$ is a subalgebra of $C^\infty(\mathbb{R}^{2n};\mathbb{R})$ under $\{\ ,\ \}$. Clearly, $\mathcal{B}$ satisfies 1) of Section 6.1. We will leave $\mathcal{A}$ as yet undetermined.

Let L be the trivial line bundle over M (there are no others). We look upon $C^\infty(M;L)$ as $C^\infty(M;\mathbb{C})$. Let

$$\theta = -\sum y_i\, dx_i \quad .$$

Then $d\theta = \omega$. Define

$$\nabla_v f = v\cdot f + 2\pi i\theta(v)f \quad .$$

Then

$$\frac{1}{2\pi i}\,\mathrm{curv}(L,\nabla) = \omega \quad .$$

We take

$$h((x,v),(x,w)) = v\bar{w} \quad ,$$

$$x \in \mathbb{R}^{2n}, \quad v,w \in \mathbb{C}$$

(we have all the data of Section 2.6). If $\varphi \in C^\infty(\mathbb{R}^{2n};\mathbb{R})$, then

$$X_\varphi = -\sum \frac{\partial\varphi}{\partial x_i}\frac{\partial}{\partial y_i} + \sum \frac{\partial\varphi}{\partial y_i}\frac{\partial}{\partial x_i} \quad .$$

Using the formula for δ (see Chapter 2), we find

$$\delta(\varphi) \quad = \quad -X_\varphi + 2\pi i\left(\sum y_i \frac{\partial\varphi}{\partial y_i} - \varphi\right) \quad .$$

The classical example is to take H_∞ to be the space of all $f \in C^\infty(\mathbb{R}^{2n})$ of the form $\psi\circ\pi$, $\pi\colon \mathbb{R}^{2n} \to \mathbb{R}^n$, $\pi(x_1,\ldots,x_n, y_1,\ldots,y_n) = (x_1,\ldots,x_n)$, $\psi \in S(\mathbb{R}^n)$. If $f = \psi\circ\pi$, $g = \eta\circ\pi$, then $\langle f,g\rangle = \langle\psi,\eta\rangle$ the $\langle\ ,\ \rangle$ on the right, the L^2 inner produce on $L^2(\mathbb{R}^n)$.

If $\varphi \in B$, then $\varphi = \psi(v,w,t)$, $v,w \in \mathbb{R}^n$, $t \in \mathbb{R}$, and

$$\psi(v,w,t)(x,y) \quad = \quad \sum v_i x_i + \sum w_i y_i + t \quad .$$

If $f \in H_\infty$ then

$$\text{1)} \qquad \delta(\psi(v,w,t))f \quad = \quad -\sum_{i=1}^n w_i \frac{\partial f}{\partial x_i} - 2\pi i\left(t + \sum_{i=1}^n v_i x_i\right) f \quad .$$

1) clearly implies $\delta(B)H_\infty \subset H_\infty$. Clearly,

$$\langle\delta(\varphi)f,g\rangle \quad = \quad \langle f,\delta(\varphi)g\rangle$$

for $f,g \in H_\infty$.

<u>Lemma 2.1</u>. Identify H (the Hilbert space completion of $(H_\infty,\langle\ ,\ \rangle)$) with $L^2(\mathbb{R}^n)$. Then $H_\infty = S(\mathbb{R}^n)$. If

$$\rho_{\psi(v,w,t)}(s) \quad = \quad S_{-2\pi}\left(s\begin{bmatrix} v \\ w \end{bmatrix}, st\right) \quad .$$

Then

$$\lim_{s\to 0} \frac{\rho_\varphi(s)f - f}{s} = \delta(\varphi)f$$

for $f \in H_\infty$ (the limit taken in the sense of $(H,< >)$).

This is not hard and we leave it to the reader. Since $S_{-2\pi}$ is irreducible, we see that I) - V) of Section 1 are satisfied for $\delta = \phi$ and $B = A$.

Now we ask: How big a Lie algebra of functions can A be maintaining I) - V)? This is an important question and has in a sense been solved (cf. Joseph [13]). We give what seems to be the maximal example.

Let A be the set of real polynomials on $\mathbb{R}^{2n}$ of degree at most two. Then A is a Lie subalgebra of $C^\infty(\mathbb{R}^{2n}; \mathbb{R})$ under Poisson bracket. Then if $g \subset A$ is the space of polynomials homogeneous of degree two, $A = g \oplus B$, g is a Lie subalgebra and B is normal. We have seen in Section 3.4 that if

$$\lambda: sp(n,\mathbb{R}) \to g$$

is defined by

$$\lambda(X)(z) = -\frac{1}{2}(z,JXz)$$

$$X \in sp(n,\mathbb{R}) \quad ,$$

$$J = \begin{bmatrix} 0 & I \\ -I & 0 \end{bmatrix}$$

(see Chapter 5 and Section 3.4), then λ is a Lie algebra isomorphism. It is not hard to check that

$$\left\{\lambda(X), \psi\left(\begin{bmatrix} v \\ w \end{bmatrix}, t\right)\right\} = \psi\left(X\begin{bmatrix} v \\ w \end{bmatrix}, t\right)$$

(Here we have written

$$\psi\left(\begin{bmatrix} v \\ w \end{bmatrix}, t\right)$$

for $\psi(v, w, t)$.)

If ϕ and ρ_ϕ exist for $\phi \in A$ satisfying I) - V) of Section 1, then we must have

$$\rho_{\lambda(X)}(s) S_{-2\pi}(u,t) \rho_{\lambda(X)}(s)^{-1} = S_{-2\pi}(\exp(sX)u, t)$$

The results of Chapter 5 now apply and Theorem 5.4.2 implies that if $W_{-2\pi}$ is the realization of W^* corresponding to $S_{-2\pi}$ then

$$\rho_{\lambda(X)}(s) = W^*(\operatorname{Exp}(sX))$$

where $\operatorname{Exp}: sp(n, \mathbb{R}) \to Mp(n, \mathbb{R})$ is the exponential map.

After a slight computation we find

a) $$\phi\left(\lambda\left(\begin{bmatrix} I & Y \\ 0 & I \end{bmatrix}\right)\right) f = -\pi i \sum_{i,j=1}^{n} (Y_{ij} x_i x_j) f \quad ,$$

$$Y = (Y_{ij}), \quad {}^tY = Y \quad ,$$

b) $$\phi\left(\lambda\left(\begin{bmatrix} Y & 0 \\ 0 & -{}^tY \end{bmatrix}\right)\right) = \frac{1}{2}\operatorname{tr}(Y)f + \sum Y_{ij}x_i \frac{\partial f}{\partial x_j}$$

$$Y = (Y_{ij}) \in Mn(\mathbb{R}) \quad ,$$

and

c) $$\phi\left(\lambda\left(\begin{bmatrix} I & 0 \\ Y & I \end{bmatrix}\right)\right)f = -\pi i \sum_{ij} Y_{ij} \frac{\partial^2}{\partial x_i \partial x_j} f$$

$$Y = (Y_{ij}), \quad {}^tY = Y \quad .$$

The theory of the metaplectic representation now implies that we have quantized (A,B) relative to (L,∇). We note that this example shows that we cannot hope to "restrict δ to A" to get ϕ satisfying I) - V). Indeed, $\delta(\varphi)$ is always a first order differential operator. Even b) above is not gotten directly from δ. Indeed,

$$\phi\left(\lambda\left(\begin{bmatrix} Y & 0 \\ 0 & -{}^tY \end{bmatrix}\right)\right) = \left(\delta\left(\lambda\left(\begin{bmatrix} Y & 0 \\ 0 & -{}^tY \end{bmatrix}\right)\right) + \frac{1}{2}\operatorname{tr} Y\right)\Big|_{H_\infty}$$

3. ISOMORPHIC QUANTIZATIONS

We return to the situation of Section 6.1. Let us call a nonzero space $H_\infty \subset C^\infty(M;L)$ with pre-Hilbert space structure

$\langle\, ,\rangle$ so that $\delta(\mathcal{B})H_\infty \subset H_\infty$ and satisfying I), II), IV), V) for $\mathcal{B} = \mathcal{A}$ a <u>polarization</u> of $(\mathcal{B},L,\nabla)$.

We say that the polarizations H_∞ and H'_∞ are <u>equivalent</u> or <u>isomorphic</u> if there is a unitary bijection $A\colon H \to H'$ so that $AH_\infty = H'_\infty$ and

$$A \circ (\delta(\varphi)|_{H_\infty}) = \delta(\varphi)|_{H'_\infty} \circ A$$

for $\varphi \in \mathcal{B}$.

The Stone-Von Neumann theorem implies that up to isomorphism the polarization described in 6.2 is the only one.

The definition of polarization given in this section is quite a bit more general than the usual definitions (e.g., Blattner [3]). In fact, we feel that it is probably too general. Since it is internal to our definition of "quantization", it contains all the drawbacks of Section 1.

We note that if H_∞ and H'_∞ are isomorphic under A and if ϕ is a quantization of $\mathcal{A}$ on H_∞ then

$$A\circ\phi(\varphi)\circ A^{-1} = \phi''(\varphi)$$

defines a quantization of $\mathcal{A}$ on H'_∞. However, there is no justification in feeling that if ϕ' is a quantization on H'_∞ then $\phi' = \phi''$ in general. However, in the example of the last section this is true that $\phi' = \phi''$.

4. THE KIRILLOV QUANTIZATION

Let G be a connected, simply connected Lie group. Let g be its Lie algebra and let g^* be the real dual space of g. We let G act on g^* as usual by

$$g \cdot f = f \circ \mathrm{Ad}(g)^{-1} .$$

Let $f \in g^*$, $\mathcal{O}_f = G \cdot f$ and ω^f the corresponding symplectic structure. We assume that ω^f is integral.

Let, for $X \in g$,

$$\lambda(X)(p) = p(X)$$

for $p \in \mathcal{O}_f$.

Let

$$\mathcal{B} = \lambda(g) \subset C^\infty(\mathcal{O}_f; \mathbb{R}) .$$

If $X \in g$ let $X^* \in \mathcal{X}(\mathcal{O}_f)$ be the induced vector field

$$X^*_p \varphi = \frac{d}{dt} \varphi(\exp tX \cdot p)|_{t=0} .$$

Then

$$\{\lambda(X), \lambda(Y)\} = \omega^f(X^*, Y^*)$$

$$= \lambda([X,Y])$$

(see Section 3.2). Thus $\mathcal{B}$ is a Lie subalgebra of $C^\infty(M;\mathbb{R})$ and clearly satisfies 1).

We recall from Section 3.5 that the condition that ω^f is integral is the same as the condition that if

$$G_f = \{g \in G \mid g\cdot f = f\}$$

then there is a character

$$\chi\colon G_f \to T^1$$

so that

$$\chi_*(X) = 2\pi i f(X)$$

$$\text{for } X \in \mathfrak{g}_f$$

the Lie algebra of G_f. Let $L \to \mathcal{O}_f$ be the line bundle constructed as in the discussion preceeding Theorem 3.5.6. Let ∇ and h also be as in that discussion.

We are now ready to attempt to quantize $\mathcal{B}$. We say that a closed subgroup H of G is <u>subordinate</u> to χ if $H \supset G_f$ and χ extends to a character of H.

We say that a subgroup P of G is a <u>polarizing subgroup</u> for χ if

$$\begin{aligned}\dim(P\cdot f) &= n\\ &= \tfrac{1}{2}\dim \mathcal{O}_f \quad .\end{aligned}$$

We assume that such a polarizing subgroup, P, exists and that G/P carries a G-invariant volume element. These are both assumptions as can be seen from general theory (see, e.g., Brezin [4], for the nonexistence of a polarizing subgroup). If G is unimodular and P is not, then there is no G-invariant measure on G/P.

We now define a space H_∞. We note that the construction of the line bundle over $\mathcal{O}_f$ corresponding to χ in Section 3.5 works for (G,P,χ). That is, let P act on the right on $G \times \mathbb{C}$ by

$$(g,z)\cdot p = (gp, \chi(p)^{-1}z) \quad .$$

Set

$$\tilde{L} = G \times \mathbb{C}/P \quad .$$

Then $\tilde{L}$ defines a line bundle over G/P. We note that there is a natural map

$$\pi\colon (G/G_f = \mathcal{O}_f) \mapsto G/P$$

by $g\cdot G_f \mapsto g\cdot P$. We define a map $\pi^!\colon L \to \tilde{L}$ as follows:

$$(g,z)\cdot G_f \mapsto (g,z)P \quad .$$

Then

$$\pi^!\colon L_x \mapsto \tilde{L}_{\pi(x)}$$

is linear.

Clearly $\pi^!: L^\times \to \tilde{L}^\times$. Hence we have a map

$$(\pi^!)^*: C^\infty(\tilde{L}^\times) \to C^\infty(L^\times) \quad .$$

We recall from Section 2.5 (see Definition 2.5.9) that if $E \overset{p}{\to} M$ is a line bundle then there is a map $C^\infty(M;E) \to C^\infty(E^\times)$ given by $S \mapsto \tilde{S}$ where $\tilde{S}(\ell)\ell = S(p(\ell))$. We also note that if $\psi(X) = \tilde{S}$ then

$$\psi(C^\infty(M;E)) = \{f \in C^\infty(E^\times) \mid f(\lambda\ell) = \lambda^{-1} f(\ell)\} \quad .$$

These observations imply that

$$(\pi^!)^*: \psi(C^\infty(\tilde{L}^\times)) \to \psi(C^\infty(L^\times)) \quad .$$

This induces a map

$$\pi_!: C^\infty(G/P;\, \tilde{L}) \to C^\infty(\mathcal{O}_f;L) \quad .$$

We take

$$H_\infty = \pi_!(C_c^\infty(G/P;\, \tilde{L})) \quad .$$

We also take

$$\langle \pi_! f, \pi_! g \rangle = \int_{G/P} \langle f(x), g(x) \rangle \, dx$$

where $\langle\ ,\ \rangle$ is the Hermitian structure on $\tilde{L}$ given by

$$\langle (g,z)\cdot P,\ (g,w)\cdot P \rangle = z\bar{w} \quad .$$

Returning to the discussion preceeding Theorem 3.5.6, we see that if $X \varepsilon g$

$$\delta(\lambda(X))S(y) = \frac{d}{dt}\Big|_{t=0} \exp(tX)\cdot S(\exp(-tX)\cdot y)$$

for $y \varepsilon \mathcal{O}_f$, $S \varepsilon C^\infty(\mathcal{O}_f; L)$.

Set

$$(\rho_{\lambda(X)}(u)S)(y) = \exp(uX)S(\exp(-uX)\cdot y)$$

for $S \varepsilon C^\infty(\mathcal{O}_f; L)$.

Set

$$(\mu_X(u)S)(y) = \exp(uX)S(\exp(-uX)y)$$

for $S \varepsilon C^\infty(G/P; L)$.

Then

$$\pi_! \circ \mu_X(u) = \rho_{\lambda(X)}(u) \circ \pi_! .$$

Hence

$$\delta(\lambda(X))\cdot H_\infty \subset H_\infty .$$

It is also clear from the above that if we take $A = B$ and $\phi(\lambda(X)) = \delta(\lambda(X))|_{H_\infty}$ then I) ,..., IX) of Section 6.1 are satisfied.

We therefore see that we will have satisfied all the conditions of quantization if V) of Section 6.1 is satisfied. We will see that it is satisfied if G is nilpotent (one of the basic theorems of Kirillov theory).

We note that if $G = H_n$, the $2n+1$ dimensional Heisenberg group and if $f \in h_n^*$ is given by $f(x,t) = t$, then $(\mathcal{O}_f, \omega^f)$ is just $(\mathbb{R}^{2n}, \omega)$,

$$\omega = \sum dx_i \wedge dy_i \quad .$$

If we take

$$P = \{((0,y),t) \mid y \in \mathbb{R}^n, \; t \in \mathbb{R}\} \quad ,$$

then the Kirillov quantization of $\mathcal{B}$ is just the classical quantization of Section 6.2.

We will see in Chapter 7 (Kirillov theory) that if G is nilpotent and $f \in g^*$ then all Kirillov quantizations of $\lambda(g)$ are isomorphic. This is a far reaching generalization of the Stone-Von Neumann theorem. However, as we shall see, the Stone-Von Neumann theorem is the main tool from analysis in the proof of the Kirillov isomorphism theorem.

We also note that if G/P does not carry a G-invariant volume, then we cannot carry out our version of the Kirillov quantization. In "most examples" G/P does not carry a G-invariant volume element. It is for this reason that the definitions of quantization (cf. Blattner [3]) are not just on a subspace of $C^{\infty}(M;L)$. In fact, it is necessary to look at $L \otimes \nu$, ν a line bundle over M which is in some sense a "square root" of the line bundle $\Lambda^{2n}T(M)^*$. One uses a tensor product connection on $L \otimes \nu$ and extends the definition of δ to $L \otimes \nu$ using this connection. We will not go into the details necessary to make these observations precise. The interested reader should consult Blattner [3] or Bernat, et al. [2].

Chapter 7

THE KIRILLOV THEORY

In this chapter we study the Kirillov theory of unitary representations of nilpotent Lie groups. We first put the classification of the irreducible unitary representations of the Heisenberg group (give in Chapter 4) into a form which at least makes sense for more general Lie groups. We then develop the necessary structure theory to show that the indicated results extend to arbitrary simply connected nilpotent Lie groups.

1. THE HEISENBERG GROUP REVISITED

In this section we take another look at the results of Section 4.4, with the intent of generalizing them to arbitrary nilpotent Lie groups.

Let H be the Heisenberg group of dimension three. That is,

$$H = \left\{\left(\begin{bmatrix} x \\ y \end{bmatrix}, t\right) \,\middle|\, x,y,t \in \mathbb{R}\right\}$$

$$\left(\begin{bmatrix} x \\ y \end{bmatrix}, t\right) \cdot \left(\begin{bmatrix} x' \\ y' \end{bmatrix}, t'\right) = \left(\begin{bmatrix} x + x' \\ y + y' \end{bmatrix}, t + t' + \frac{1}{2}(xy' - x'y)\right) .$$

The Lie algebra of H, h, is also the set of all pairs $\left(\begin{bmatrix} x \\ y \end{bmatrix}, t\right)$, $x,y,t \in \mathbb{R}$ with

$$\left[\left(\begin{bmatrix}x\\y\end{bmatrix},t\right),\left(\begin{bmatrix}x'\\y'\end{bmatrix},t'\right)\right] = (0,\ xy'-yx') \quad .$$

We have already observed that $\exp\colon h \to H$ is the identity map.

If $f \in h^*$, we study $\mathcal{O}_f = H\cdot f$, $(h\cdot f = f\circ \mathrm{Ad}(h)^{-1})$. We use the notation

$$\beta\left(\begin{bmatrix}x\\y\end{bmatrix},\begin{bmatrix}x'\\y'\end{bmatrix}\right) = xy' - yx' \quad .$$

We note that

$$(x,t)\cdot(y,s)(-x,-t) = (y,\ x+\beta(x,y)) \quad .$$

Hence we have

1) $$\mathrm{Ad}((x,t))\cdot(y,s) = (y,\ s+\beta(x,y)).$$

Using 1) we find

2) $$((x,t)\cdot f)(y,s) = f(y,s) - \beta(x,y)f(0,1) \quad .$$

Using 2) we can completely describe the orbits in h^*.

a) $f(0,1) = 0$, then 2) implies $h\cdot f = f$, $h \in H$.

b) $f(0,1) \neq 0$. Set $f(0,1) = \mu$. Then

$$f(x,t) = \lambda_f(x) + \mu t \quad .$$

Since $\lambda_f \in (\mathbb{R}^2)^*$, there exists $y_f \in \mathbb{R}^2$ so that

$$\left(\left(\frac{1}{\mu} y_f, 0\right) \cdot f\right)(x,t) = \mu t \quad .$$

Let for $\mu \varepsilon \mathbb{R}$, $\mu \neq 0$, $h_\mu \varepsilon h^*$ be defined by

$$h_\mu^* (x,t) = \mu t \quad .$$

Then we have seen that if $f(0,1) \neq 0$, then $f \varepsilon H \cdot h_{f(0,1)}^*$.

We collect this material into a lemma.

Lemma 1.1. There are two types of orbits in h^*.

1) $\mathcal{O}_f = \{f\}$ if $f(0,1) = 0$.

2) $\mathcal{O}_\mu = \{f \varepsilon h^* | \, f(0,1) = \mu\}$, $\mu \neq 0$, $\mu \varepsilon \mathbb{R}$.

According to the results of Section 4.4, Lemma 1.1 implies that there is a one-to-one correspondence between the orbits in h^* and the equivalence classes irreducible unitary representation of H. Indeed, we correspond $\mathcal{O}_f$, $f(0,1) = 0$ to

$$T_f((x,t)) = e^{2\pi i f(x,t)} \quad .$$

Then $(T_f, \mathbb{C})$ is the most typical one dimensional unitary representation of H. To $\mathcal{O}_\mu$ we associate $(S_{2\pi\mu}, L^2(\mathbb{R}))$ (see Section 4.4). In order to generalize this correspondence we must decide on a more natural method of establishing this correspondence.

Our first method of establishing the correspondence will be through characters. If G is a Lie group and (π, V) is a finite dimensional representation of G then the <u>character</u> χ_π of π is defined by

$$\text{I)} \qquad \chi_\pi(g) = \operatorname{tr} \pi(g) \quad .$$

If (π, V) is an infinite dimensional unitary representation of G then I) makes no sense. We reformulate the definition of χ_π in a way that makes sense for infinite dimensional unitary representations. Suppose that (π, V) is a finite dimensional representation of G. If $f \in C_c^\infty(G)$ define

$$\pi(f) = \int_G f(g)\pi(g)\, dg \quad .$$

(dg is always a left invariant measure on G.)

Set

$$\Theta_\pi(f) = \operatorname{tr}(\pi(f)) \quad .$$

It is a direct computation to see that

$$\text{II)} \qquad \Theta_\pi(f) = \int_G f(g)\chi_\pi(g)\, dg$$

$$f \in C_c^\infty(G) \quad .$$

Clearly II) says that Θ_π contains all the information of χ_π. Definition II) does generalize to infinite dimensions.

We first record a simple lemma whose proof we will leave to the reader (cf. Wallach [26], Chapter 2).

Lemma 1.2. Let (π,V) be a unitary representation of G. If $f \in C_c^\infty(G)$ then there is a unique bounded operator $\pi(f)$ on V so that

$$\langle \pi(f)v,w\rangle = \int_G f(g)\langle \pi(g)v,w\rangle \; dg$$

Definition 1.3. A unitary representation (π,V) of G is said to be of trace class if $\pi(f)$ is a trace class operator for every $f \in C_c^\infty(G)$. (See Appendix I to this chapter for a discussion of trace class operators.) If (π,V) is of trace class, then we define the character of π to be

$$\operatorname{tr} \pi(f) = \Theta_\pi(f)$$

$$\text{for } f \in C_c^\infty(G) \quad .$$

Using the properties of the trace (see the Appendix to this chapter) we see that if (π,V) and (σ,W) are unitarily equivalent then $\Theta_\pi = \Theta_\sigma$ in the sense that π is trace class if and only if σ is trace class and the characters are the same.

Let $\hat{G}_{tr}$ denote the set of equivalence classes of irreducible trace class unitary representations of G. If $\omega \in \hat{G}_{tr}$ define $\Theta_\omega = \Theta_\pi$ if $\pi \in \omega$.

We now return to H. We first give a choice of invariant measure on H. Using the formula for multiplication in H it is easily seen that

$$dx \wedge dy \wedge dt = d\left(\begin{bmatrix} x \\ y \end{bmatrix}, t\right)$$

is both left and right invariant (e.g., if $f \in C_c^\infty(H)$ then

$$\int_{\mathbb{R}^3} f\left(\left(\begin{bmatrix} x' \\ y' \end{bmatrix}, t'\right)\left(\begin{bmatrix} x \\ y \end{bmatrix}, t\right)\right) dx\,dy\,dt = \int_{\mathbb{R}^3} f\left(\begin{bmatrix} x+x' \\ y+y' \end{bmatrix}, t + t' + \frac{1}{2}\beta\left(\begin{bmatrix} x' \\ y' \end{bmatrix}, \begin{bmatrix} x \\ y \end{bmatrix}\right)\right) dx\,dy\,dt$$

Since dx, dy and dt are translation invariant, we see that $dx \wedge dy \wedge dt$ is left invariant, etc.)

Now that we have dh we are ready to compute the Θ_ω, $\omega \in \hat{H}_{tr}$. We first look at the one dimensional unitary representations. Let $g \in h^*$ be such that

$$h \cdot g = g$$

$$\text{for all } h \in H .$$

Then

$$\pi_g(h) = e^{2\pi i g(h)}$$

defines a unitary representation of H. (Here we have identified H as a space with h since $\exp: h \to H$ is the identity.) Let $f \in C_c^\infty(H)$. Then

$$\pi_g(f) = \int_{h} f(X)e^{2\ ig(X)}\ dX$$

(here we use dX for $dx \wedge dy \wedge dt$, $X = \left(\begin{bmatrix} x \\ y \end{bmatrix}, t\right)$).

Define

$$\tilde{f}(g) = \int_{h} f(X)\ e^{2\pi i g(X)}\ dX$$

for $f \in S(h)$

(the Schwartz space of $h = \mathbb{R}^3$), $g \in h^*$. Then we have shown that if $g \in h^*$, $h \cdot g = g$ for $h \in H$ then

$$\text{III)} \qquad \Theta_{\pi_g}(f) = \tilde{f}(g) = \tilde{f}(O_g)$$

Let us look at $S_{2\pi\mu} = \pi_O$, $O = O_\mu = \{g \in h^* |\ g(0,1) = \mu\}$. Let $f \in S(h)$. If $\varphi \in S(\mathbb{R})$, then

$$(\pi_\mu(f)\varphi)(z) = \int_{\mathbb{R}^3} f\left(\begin{bmatrix} x \\ y \end{bmatrix}, t\right) \exp(2\pi i\mu(t + x(z - \tfrac{1}{2}y))\varphi(z - y)\ dx\, dy\, dt$$

$$= \int_{\mathbb{R}^3} f\left(\begin{bmatrix} x \\ y - z \end{bmatrix}, t\right) \exp(2\pi i\mu(t + \tfrac{1}{2}x(z+y)))\varphi(y)\ dx\, dy\, dt$$

Set

$$K_f(z,y) = \int_{\mathbb{R}^3} f\left(\begin{bmatrix} x \\ y - z \end{bmatrix}, t\right) \exp(2\pi i\mu(t + \tfrac{1}{2}x(z+y))\ dx\ dt \qquad .$$

Then

$$(\pi_{\mathcal{O}}(f)\varphi)(z) = \int_{-\infty}^{\infty} K_f(z,y)\varphi(y)\,dy \quad .$$

Clearly $K_f \in \mathcal{S}(\mathbb{R}^2)$. Applying Lemma A.7.I of the Appendix to this chapter we find that $\pi_{\mathcal{O}}(f)$ is trace class and

$$\operatorname{tr} \pi_{\mathcal{O}}(f) = \int_{-\infty}^{\infty} K_f(x,x)\,dx \quad .$$

To compute the above integral we note

Lemma 1.4. If $u \in \mathcal{S}(\mathbb{R})$ then

$$\int_{-\infty}^{\infty}\int_{-\infty}^{\infty} u(x)e^{2\pi ixy}\,dx\,dy = u(0) \quad .$$

Proof.

$$\int_{-\infty}^{\infty} u(x)e^{2\pi ixy}\,dx = \sqrt{2\pi}\,\hat{u}(-2\pi y) \quad .$$

$$\sqrt{2\pi}\int_{-\infty}^{\infty} \hat{u}(-2\pi y)\,dy = \frac{1}{\sqrt{2\pi}}\int_{-\infty}^{\infty} \hat{u}(-y)\,dy$$

$$= u(0) \quad .$$

(Here we have used the Fourier integral theoreom. See Section 4.1.)

Using Lemma 1.4 we find (using obvious changes of variables)

$$\text{IV)} \qquad \operatorname{tr} \pi_{\mathcal{O}}(f) = \frac{1}{|\mu|} \int_{-\infty}^{\infty} f(0,t)\, e^{2\pi i \mu t}\, dt \quad .$$

We now come to the critical observation: $\mathcal{O}_\mu = H \cdot h^*_\mu$ is a symplectic manifold with a canonical symplectic structure $\omega^{h^*_\mu} = \omega^\mu$. We note that

$$H_{h^*_\mu} = \{h \in H \mid h \cdot h^*_\mu = h^*_\mu\} \quad .$$

Using 2) above

$$(x,t) \cdot h^*_\mu = h^*_\mu - \mu\beta(x,\cdot) \quad .$$

Thus we see

$$3) \qquad H_{h^*_\mu} = \{(0,t) \mid t \in \mathbb{R}\} = z$$

Thus we have a bijective map $\psi\colon H/z \to \mathcal{O}_\mu$, $\psi(hz) = h \cdot h^*_\mu$. Furthermore, if $\eta(x,0) = (x,0) \cdot h^*_\mu$, then $\eta\colon \mathbb{R}^2 \to \mathcal{O}_\mu$ is a bijection. If $X = (x,0)$, $Y = (y,0)$ are chosen in h then our formula for ω^μ gives

$$\begin{aligned} \omega^\mu_p(X^*, Y^*) &= p([(x,0),(y,0)]) \\ &= p(0, \beta(x,y)) \\ &= \mu\beta(x,y) \quad . \end{aligned}$$

Hence we see

4) $$\eta^*\omega^\mu = \mu dx \wedge dy$$

In this case, $\dim \mathcal{O}^\mu = 2$. Hence, $\omega^\mu = \Omega^\mu$.

<u>Lemma 1.5</u>. If $\mu \neq 0$ then

$$\int_{\mathcal{O}_\mu} (\tilde{f}|_{\mathcal{O}_\mu})\Omega^\mu = \frac{1}{|\mu|} \int_{-\infty}^{\infty} f(0,t)e^{2\pi i\mu t}\, dt \quad .$$

<u>Proof</u>.

$$\int_{\mathcal{O}_\mu} (f|_{\mathcal{O}_\mu})\Omega^\mu = |\mu| \int_{\mathbb{R}^2} \tilde{f}((x,0)\cdot h^*\mu)\, dx\, dy$$

$$= |\mu| \int_{\mathbb{R}^2} \tilde{f}(\mu(h_1 - x^\#))\, dx \quad ,$$

$$x^\#(y) = \beta(x,y) \quad , \quad x^\# \in \mathbb{R}^2 \quad .$$

Since $(\mu x)^\# = \mu x^\#$, we find

$$\int_{\mathcal{O}_\mu} (f|_{\mathcal{O}_\mu}) = \frac{1}{|\mu|} \int_{\mathbb{R}^2} \tilde{f}(\mu h_1^* + x^\#)\, dx \quad .$$

Now arguing as in Lemma 1.4 we find (using the Fourier integral theorem) that

$$\int_{\mathbb{R}^2} \tilde{f}(\mu h_1^* + x^\#)\, dx = \int_{-\infty}^{\infty} f(0,t)e^{2\pi i\mu t}\, dt$$

Q.E.D.

Using the definition

$$\int_{\{f\}} h\Omega^f = h(f)$$

$$\text{for } f \in \mathfrak{h}^*, \quad H\cdot f = \{f\} \quad ,$$

we have shown

<u>Theorem 1.6</u>. If $\mathcal{O} \subset \mathfrak{h}^*$ is an orbit for the action of H on $\mathfrak{h}^*$, then

1) $\pi_{\mathcal{O}}$ is trace class

2) If $f \in \mathcal{S}(\mathfrak{h})$, then

$$\Theta_{\pi_{\mathcal{O}}}(f) = \int_{\mathcal{O}} (\tilde{f}|_{\mathcal{O}})\Omega^{\mathcal{O}}$$

(here $\Omega^{\mathcal{O}} = \Omega^h$, $h \in \mathcal{O}$).

Theorem 1.6 is certainly a more intrinsic correspondence between the orbits in $\mathfrak{h}^*$ and the equivalence classes of irreducible representations of H. Also $\int_{\mathcal{O}}(\tilde{f}|_{\mathcal{O}})\Omega^{\mathcal{O}}$ has an obvious generalization to <u>arbitrary</u> Lie groups. There is still an ingredient missing in our correspondence between orbits and elements of $\hat{H}$. We have used the classification theorem of Section 4.4 to assign a representation to each orbit. We must find another technique to find this representation $\pi_{\mathcal{O}}$ which would generalize to "arbitrary" Lie groups. The way to do this comes from the "Kirillov quantization" of Section 6.4.

Definition 1.7. Let $\mathfrak{g}$ be a Lie algebra and let $f \in \mathfrak{g}^*$. Then a polarizing subalgebra of $\mathfrak{g}$ for f (or simply a polarization of f) is a subalgebra $\mathfrak{h}$ of $\mathfrak{g}$ satisfying

1) $f([\mathfrak{h},\mathfrak{h}]) = 0$

2) If $V \subset \mathfrak{g}$ is a subspace so that $f([X,Y]) = 0$ for $X,Y \in V$ and $V \supset \mathfrak{h}$ then $V = \mathfrak{h}$.

It is not true that there exist polarizations for general $\mathfrak{g}$ and f. We will see, however, that if $\mathfrak{g}$ is nilpotent (see Section 7.2), then there exist polarizations for arbitrary $f \in \mathfrak{g}^*$.

Let us see what Definition 1.7 means for $\mathfrak{g}$ the three dimensional Heisenberg Lie algebra. Let $f \in \mathfrak{g}^*$

a) $f(0,1) = 0$. Then $\mathfrak{h} = \mathfrak{g}$ is the unique polarization. Indeed, $[\mathfrak{g},\mathfrak{g}] = \{(0,t) \mid t \in \mathbb{R}\} = \mathfrak{z}$.

b) $f(0,1) \neq 0$. If $V \subset \mathfrak{h}$ is a subspace so that $f([X,Y]) = 0$ for $X,Y \in V$, then $V + \mathfrak{z} = V'$ still satisfies $f([X,Y]) = 0$, $X,Y \in V'$. But then a subspace $W \subset \mathfrak{h}$ such that

i) $f([X,Y]) = 0$, $X,Y \in W$

ii) W is maximal relative to i)

is automatically a subalgebra (in fact, a normal subalgebra).

Let W satisfy i), ii). Then $W = U \oplus z$, $U \subset \mathbb{R}^2 \times \{0\}$. If $X = (x,0)$, $Y = (y,0)$, $x,y \in \mathbb{R}^2$, then

$$f([X,Y]) = f(0,1)\beta(x,y) \quad .$$

This implies that $U \subset \mathbb{R}^2$ is a one dimensional subspace. Hence, in this case every polarization of f is of the form $U \times \mathbb{R}$, $U \subset \mathbb{R}^2$, $\dim U = 1$.

Returning to the general case g, $f \in g^*$ and h g a polarization of f. Let G be the connected simply connected Lie group with Lie algebra g. Let $H \subset G$ be a closed subgroup such that

1) H is maximal subject to the condition that the Lie algebra of H is h.

2) There is a unitary character $\chi\colon H \to T^1$ so that χ has differential $2\pi i f|_h$.

3) $H \supset G_f$.

Lemma 1.8. If 1), 2), 3) are satisfied, then ω^f on $\mathcal{O}_f$ is integral.

Proof. See Chapter 3.

We therefore know that if H exists, then ω^f must be integral. We will also assume that H is connected.

Definition 1.9. If $f \in g^*$, H exists and is connected and if G/H has G-invariant measure define

$$\pi(f,h) = \mathrm{Ind}_H^G(\chi_f) ,$$

χ_f is the character on H whose differential is $2\pi if$. (See Appendix II to this chapter for the definition of $\mathrm{Ind}_H^G(\chi_f)$.)

We return to the case where H is the three dimensional Heisenberg group. h is the Lie algebra of H.

Lemma 1.10. If $f \in h^*$ and V is a polarization of f, then $\pi(f,V)$ is unitarily equivalent with $\pi(\mathcal{O}_f)$.

Proof. If $H \cdot f = \{f\}$ then the result is clear. We therefore assume that

$$\mathcal{O}_f = \mathcal{O}_\mu = \{g \in h^* | \; g(0,1) = \mu\} , \qquad \mu \neq 0 .$$

We must show that $\pi(f,V)$ is unitarily equivalent with $S_{2\pi\mu}$. We identify $V = U \times \mathbb{R} = h$, $\dim U = 1$ with $V \subset H$ as a Lie group (they are both abelian). Let

$$U = \mathbb{R}v , \qquad v \neq 0$$

($\dim U = 1$). Let $v' \in \mathbb{R}^2$ be defined by

$$\beta(v',v) = -1 .$$

Then if

$$W = (\mathbb{R}v', 0)$$

it is clear that

$$W \cdot V = H$$

and the map $W \times V \to H$ given by multiplication is a diffeomorphism. We may thus identify H/V with W. It is easy to see that we may take H-invariant measure on H/V to be $dz = d((zv',0))$.

Now the representation space for $\pi(f,v)$ is the space $\mathcal{H}_{f,V}$ of all $\varphi\colon H \to \mathbb{C}$ such that (see Appendix II):

1) $\varphi(hv) = \chi_f(v)^{-1}\varphi(h)$

$v \in V, \quad h \in H$

2) $\displaystyle\int_{H/V} |\varphi(h)|^2 \, d(hV) < \infty .$

Also

$$(\pi(f,V)(g)\,)(x) = \varphi(g^{-1}x) .$$

If $f' = g\cdot f, \quad g \in H$ we define

$$Ag\colon \mathcal{H}_{f,V} \to \mathcal{H}_{f',V}$$

by

$$A_g\varphi(h) = \varphi(hg) .$$

Then

$$\begin{aligned}(A_g\varphi)(hv) &= \varphi(hvg)\\ &= \varphi(hgg^{-1}vg)\\ &= e^{-2\pi i f(g^{-1}vg)}(A_g\varphi)(h)\\ &= e^{-2\pi i f'(v)}(A_g\varphi)(h) \quad .\end{aligned}$$

Hence 1) is satisfied for f' and $A_g\varphi$. 2) is also easy. We leave it to the reader to check that A_g is unitary. Thus we need only prove that $\pi(h^*_\mu, V)$ is unitarily equivalent with $S_{2\pi\mu}$.

If φ satisfies 1) $(f = h^*_\mu)$ then set

$$(B\varphi)(z) = \varphi((zv',0)) \quad .$$

A simple computation gives, for $\varphi \in L^2(\mathbb{R})$,

$$(B\pi(h^*_\mu,V)(xv + yv',t)B^{-1}\varphi)(z) = e^{2\pi i\mu(t+x(z-\frac{1}{2}y))}\,\varphi(z - y) \quad .$$

Hence B gives the unitary equivalence. Q.E.D.

There is one more fact that we may observe in this case. We saw that

$$\Theta_{\pi(\mathcal{O}_\mu)}(f) = \frac{1}{|\mu|}\int_{-\infty}^{\infty} f((0,t))e^{2\pi i\mu t}\,dt \quad ,$$

for $f \in \mathcal{S}(\mathfrak{h})$.

From this we can conclude that

$$f(0) = \int_{\mathbb{R}-\{0\}} \Theta_{\pi(\mathcal{O}_\mu)}(f)|\mu|\, d\mu \quad .$$

We have now derived the full Kirillov theorem for the Heisenberg group.

<u>Theorem 1.11</u>. Let H be the three dimensional Heisenberg group with Lie algebra h.

1) If $f \in h^*$ then there is a polarization V of f.

2) $\pi(f,V)$ is irreducible and if $f' \in h^*$, V' is a polarization of f' then $\pi(f',V')$ is equivalent with $\pi(f,V)$ if and only if $\mathcal{O}_f = \mathcal{O}_{f'}$.

3) Set for $f \in h^*$, V a polarization of f the equivalence class of $\pi(f,V)$ equal to $\pi(\mathcal{O}_f)$ (this is well defined by 2)). Then $\pi(\mathcal{O})$ is trace class for each orbit in h^* and if Ω is the Pfaffian of $\omega^{\mathcal{O}}$ ($\Omega = \omega^{\mathcal{O}}$ in this case)

$$\Theta_{\pi(\mathcal{O})}(\varphi) = \int_{\mathcal{O}} (\tilde{\varphi}|_{\mathcal{O}})\Omega \quad ,$$

$\varphi \in S(h)$

($\tilde{\varphi}(g) = \int_h e^{2\pi i g(X)}\, \varphi(X)\, dX$ with dX given as in this section.)

4) If $\varphi \varepsilon\, S(h) = S(\mathrm{H})$, then

$$\varphi(0) = \int_{\mathbb{R}-\{0\}} \Theta_{\pi(\mathcal{O}_\mu)}(\varphi)\,|\mu|\, d\mu \quad .$$

We shall see that every part of Theorem 1.11 generalizes to arbitrary simply connected Lie groups. The rest of this chapter will be devoted to the proof of this assertion. The next section develops the pertinent definitions and structure theory for nilpotent Lie groups.

2. NILPOTENT LIE GROUPS

Let g be a Lie algebra over $\mathbb{R}$.

Definition 2.1. g is said to be nilpotent if, for any $X \varepsilon g$ there is a k so that $(\mathrm{ad}\ X)^k = 0$ (recall $\mathrm{ad}\ X \cdot Y = [X,Y]$).

Lemma 2.2. If $A \subset M_n(\mathbb{R})$ is a Lie subalgebra such that if $X \varepsilon A$ then there is k so that $X^k = 0$, then there is $v \varepsilon \mathbb{R}^n$, $v \neq 0$ so that $Xv = 0$ for all $X \varepsilon A$.

Proof. By induction on $\dim A$. If $\dim A = 1$, then the result is clear. Suppose that the result is true for $1 \leq \dim A \leq r$. We assume that $\dim A = r$.

Let $A_1 \subset A$ be a proper subalgebra of maximal dimension. Let A_1 act on A/A_1 by

$$\widetilde{\text{ad}}(X)\cdot(Y+A_1) = [X,Y] + A_1$$

$$Y \in A, \quad X \in A_1 \quad .$$

Set

$$Y + A_1 = \overline{Y}$$

$$\text{for} \quad Y \in A \quad .$$

By the inductive hypothesis there is $\overline{X} \in A/A_1$, $\overline{X} \neq 0$, so that

$$\widetilde{\text{ad}}(Y)\cdot\overline{X} = 0 \quad ,$$

$$Y \in A_1 \quad .$$

But then $\mathbb{R}X + A_1$ is a subalgebra of A containing A_1. Hence $A = \mathbb{R}X \oplus A_1$. Clearly, $\text{ad}\, X\cdot A_1 \subset A_1$. Hence $[A,A_1] \subset A_1$.

Let

$$V_1 = \{v \in \mathbb{R}^n \mid Y\cdot v = 0,\ Y \in A_1\} \quad .$$

Then $\dim V_1 \geq 1$. If $v \in V_1$, $Y \in A_1$, then

$$\begin{aligned} YXv &= [Y,X]v + XYv \\ &= 0 \quad . \end{aligned}$$

Hence, $XV_1 \subset V_1$. Since $X^k = 0$ for some k, there is $v_0 \in V_1$, $v_0 \neq 0$ so that $Xv_0 = 0$. But then $Z\cdot v_0 = 0$ for all $Z \in A$.

Q.E.D.

Corollary 2.3. Let g be a Lie algebra over $\mathbb{R}$. Then the following are equivalent:

1) g is nilpotent

2) There exist ideals $g_i \subset g$ so that $\dim g_i = i$, $i = 0,\ldots,\dim g$, and $[g,g_i] \subset g_{i-1}$.

Proof. Clearly 2) implies 1). To prove that 1) implies 2), let $A = \operatorname{ad} g \subset L(g,g)$. Let $X_1 \neq 0$ be such that $A \cdot X_1 = 0$ (see Lemma 2.1). Set $g_1 = \mathbb{R}X_1$. g acts on g/g_1 by

$$\widetilde{\operatorname{ad}}_1(X)(Y+g_1) = [X,Y] + g_1 .$$

There is $X_2 \notin g_1$ so that

$$\widetilde{\operatorname{ad}}_1(X)(X+g_1) = 0$$

if $\dim g > 1$. Set

$$g_2 = \mathbb{R}X_2 + \mathbb{R}X_1 ,$$

etc.

Lemma 2.4. Let G be a Lie group with nilpotent Lie algebra g. Let (ρ,V) be a finite dimensional representation of G so that $\rho(X)$ is nilpotent for $X \in g$. If $v \in V$ and $X \in g$ is such that $\rho(\exp X)\cdot v = v$, then $\rho(X)v = 0$.

Proof. Let $\lambda_1,\ldots,\lambda_n$ be a basis of V^*. Set

$$f_i(t) = \lambda_i(\rho(\exp tX)v - v) \quad .$$

Then

$$f_i(t) = \sum_{k\leq N} \frac{t^k}{k!} \lambda_i(\rho(X)^k v) - \lambda_i(v) \quad ,$$

$(\rho(X)^N = 0)$. Thus f_i is a polynomial in t for $i = 1,\ldots,n$. But $f_i(n) = 0$, $n \in \mathbb{Z}$, $n \geq 0$. Hence $f_i = 0$. This implies

$$\rho(\exp tX)v = v$$

for all $t \in \mathbb{R}$.

Hence

$$\rho(X)v = \frac{d}{dt}\rho(\exp tX)v|_{t=0} = 0$$

Q.E.D.

Definition 2.5. A connected Lie group is said to be nilpotent if it has a nilpotent Lie algebra.

Lemma 2.6. Let G be a simply connected, nilpotent Lie group with Lie algebra g. Let $h \subset g$ be a subalgebra and let $X_1,\ldots,X_k \in g$ be linearly independent so that

1) $\sum_{i=1}^{k} \mathbb{R}X_i \oplus h = g$

2) If $g_i = \sum_{j\geq i} \mathbb{R}X_j + h$ then g_i is a subalgebra.

Then the map

$$\Phi: \mathbb{R}^k \times h \to G$$

given by

$$\Phi(t_1,\ldots,t_k,X) = \exp t_1X_1 \cdots \exp t_kX_k \cdot \exp X$$

is a surjective diffeomorphism.

Furthermore, such a basis $X_1,\ldots,X_k$ for g/h always exists.

Proof. We first show that $\{X_1,\ldots,X_k\}$ exists. Let

$$k = \dim g/h .$$

Then Lemma 2.2 implies that there is $\overline{X}_k \in g/h$ so that

$$[\overline{h,X_k}] = 0 , \qquad \overline{X}_k \neq 0 .$$

Let

$$g_k = \mathbb{R}X_k + h .$$

Replace h with g_k and continue.

We first prove a special case of the lemma.

a) If G is a simply connected Lie group with Lie algebra g, then $\exp: g \to G$ is a diffeomorphism. (This is the case $\dim h = \dim g$ or $k = 0$ in the lemma.)

We prove a) by induction on the dimension of $\mathfrak{g}$. If $\dim \mathfrak{g} = 1$ the result is clear. Suppose true for $1 \leq \dim \mathfrak{g} \leq k$ and that $\dim \mathfrak{g} = k+1$. Let $\mathfrak{z}$ be the center of $\mathfrak{g}$. Then $\mathfrak{z} \neq 0$ by Lemma 2.3. Set

$$Z = \exp \mathfrak{z} \quad .$$

Then G/Z is a connected, simply connected, Lie group with Lie algebra $\mathfrak{g}/\mathfrak{z}$. The inductive hypothesis implies that $\exp\colon \mathfrak{g}/\mathfrak{z} \to G/Z$ is a diffeomorphism.

Let $\psi\colon G \to G/Z$ be the canonical projection. $\psi\colon \mathfrak{g} \to \mathfrak{g}/\mathfrak{z}$ its differential $(\psi(g) = gZ,\ \psi(X) = X + \mathfrak{z})$. Then

$$\psi(\exp X) = \exp(\psi(X)) \quad .$$

Let $\mathfrak{g} = V \oplus \mathfrak{z}$, V a subspace. Then

$$\psi(\exp X) = \exp \psi(X) \quad ,$$

$$X \in V \quad .$$

Hence $\exp\colon V \to \exp V$ is a diffeomorphism onto its image.

Now if $X_i \in V$, $z_i \in \mathfrak{z}$, $i = 1,2$ and

$$\exp(X_1 + Z_1) = \exp(X_2 + Z_2)$$

then since

$$\exp(X_i + Z_i) = \exp(X_i)\exp(Z_i)$$

$([X_i,Z_i] = 0)$ we see that

$$\psi(\exp X_1) = \psi(\exp X_2) .$$

Hence $X_1 = X_2$. Since $\psi(\exp V) = G/Z$ we see that $G = \exp V\cdot Z$. Thus $\exp\colon g \to G$ is surjective. By the above, if $u(X,z) = \exp Xz$, $X \in V$, $z \in Z$, then u is bijective. By computing differentials one sees easily that u is open. Hence u is a diffeomorphism. Since G is simply connected, this implies that Z is simply connected. But then $\exp\colon g \to G$ is a diffeomorphism. This proves a).

b) If we prove the result for $k = 1$, the result is true for all k.

Indeed, we know the result for $k = 0$ (a)). Let

$$G_2 = \exp g_2 .$$

Then

$$\psi\colon \mathbb{R} \times g_2 \to G ,$$

$$\psi(t,X) = \exp tX_1 \exp X$$

is a diffeomorphism. By a) $\exp\colon g_2 \to G_2$ is a covering map. If $X,Y \in g_2$ and $\exp X = \exp Y$ then $\exp X = \exp Y$, since $\exp\colon g \to G$ is injective. Hence $\exp\colon g_2 \to G_2$ is a diffeomorphism. Now use the case $k = 1$ for X_2, g_3, etc.

We prove the result for $k = 1$. Let $H = \exp h$. $\exp\colon h \to H$ is a diffeomorphism. H is a normal subgroup of G. Hence

$$G = \exp \mathbb{R}X_1 \cdot H \quad .$$

If $\psi(t,X) = \psi(t',X')$, then $\exp(t-t')X \in H$. But then $(t-t')X \in \mathfrak{h}$. Since $\exp: \mathfrak{g} \to G$ is injective. Thus $t = t'$. But then $X = X'$. Thus ψ is bijective. Let

$$\tilde{\psi}: \mathbb{R} \times H \to G \quad ,$$

$$\tilde{\psi}(t,h) = \exp tX \cdot h \quad .$$

Then

$$\tilde{\psi}_{*(t,h)}\left(X \frac{d}{dt} , Z\right) = (\lambda \operatorname{Ad}(h)^{-1}X + Z)_{\exp tX \cdot h} \quad .$$

Hence ψ is regular at each (t,h). Thus ψ is a diffeomorphism. Q.E.D.

3. UNIPOTENT REPRESENTATION OF NILPOTENT LIE GROUPS

Let G be a nilpotent Lie group with Lie algebra $\mathfrak{g}$. A finite dimensional representation (ρ,V) (V a real vector space) is said to be <u>unipotent</u> if, for each $g \in G$,

$$(\rho(g) - I)^k = 0$$

$$\text{for some } k > 0, \ k \in \mathbb{Z} \quad .$$

If (ρ,V) is a unipotent representation of G and if (ρ,V) also denotes the corresponding representation of $\mathfrak{g}$, and if $X \in \mathfrak{g}$, then there is a $k \in \mathbb{Z}$, $k > 0$ so that

$\rho(X)^k = 0$. Indeed,

$$\rho(\exp tX) - I = t\left(\rho(X) + t\sum_{\ell=2}^{\infty} \frac{t^{\ell-2}}{\ell!}\rho(X)^\ell\right) .$$

Since $(\rho(\exp tX) - I)^k = 0$ we see then

$$0 = t^k\left(\rho(X) + t\sum_{\ell=2}^{\infty} \frac{t^{\ell-2}}{\ell!}\rho(X)^\ell\right)^k$$

$$= t^k\rho(X)^k + t^k(\text{higher powers of } t) .$$

Equating powers of t gives $\rho(X)^k = 0$.

Lemma 2.2 implies that

$$V = V_0 \supset V_1 \supset V_2 \supset \cdots \supset V_n = 0$$

with $\dim V_i = i$ and $\rho(\mathfrak{g})V_i \subset V_{i-1}$. (See the proof of Corollary 2.3.)

Set, for $v \in V$,

$$\mathfrak{g}_v = \{X \in \mathfrak{g} \mid \rho(X)v = 0\}$$

and

$$G_v = \{g \in G \mid \rho(g)\cdot v = v\} .$$

Then Lemma 2.4 implies that

$$G_v = \exp \mathfrak{g}_v .$$

Fix $v \in V$, $v \neq 0$. Let $\pi_j : V \to V/V_j$ be the canonical projection. Set

$$\mathfrak{g}_j = \{X \in \mathfrak{g} \mid \rho(X)v \in V_j\}$$

Then

$$\mathfrak{g} = \mathfrak{g}_0 \supset \mathfrak{g}_1 \supset \mathfrak{g}_2 \supset \cdots \supset \mathfrak{g}_k = \mathfrak{g}_v$$

Let

$$1 \leq j_1 < j_2 < \cdots < j_k \leq n$$

be the indices such that

$$\mathfrak{g}_{j_i} \neq \mathfrak{g}_{j_i - 1} \quad .$$

Let $e_1,\ldots,e_n$ be a basis of V so that

$$V_i = \sum_{j>i} \mathbb{R}e_j \quad .$$

Then

$$V_i = \mathbb{R}e_i \oplus V_{i+1} \quad .$$

Let

$$X_i \in \mathfrak{g}_{j_i - 1}$$

be such that

$$\pi_{j_i}(\rho(X_i)v) = \pi_{j_i}(e_{j_i}) \quad .$$

(This is possible since

$$\dim \pi_{j_i}(\rho(g_{j_i-1})v) = 1 \quad .)$$

Lemma 2.6 implies that the map

$$\mathbb{R}^k \times G_v \to G \quad ,$$

$$(t_1,\ldots,t_k,g) \mapsto \exp t_1X_1 \cdots \exp t_kX_kg$$

is a surjective diffeomorphism.

Since ρ is a unipotent representation of G we see that (we are using $\rho(X_i)^{k_i} = 0$)

$$\rho(\exp t_1X_1 \cdots \exp t_kX_k)v = \sum_{j=1}^{n} p_j(t_1,\ldots,t_k)e_j$$

with p_j a polynomial in $t_1,\ldots,t_k$.

1) p_j is really a function only of the variables $t_1,\ldots,t_r$ where $r = \max\{j_i \leq j\}$.

To see this we note that if $j_i > j$, then $X_i \in g_j$. Thus $\rho(X_j)v \in V_{j+1}$. Thus

$$\pi_j(\rho(\exp t_1X_1 \cdots \exp t_kX_k)v) = \pi_j(\rho(\exp t_1X_1 \cdots \exp t_rX_r)v)$$

This implies 1).

Now,

$$\pi_{j_i}(\rho(\exp t_1X_1\cdots\exp t_kX_k)v) = \pi_{j_i}(\rho(\exp(t_1X_1\cdots\exp t_iX_i)v)$$

$$= \pi_{j_i}(\rho(\exp t_1X_1\cdots\exp t_{i-1}X_{i-1})$$

$$\circ(v + t_ie_{j_i}))$$

Hence

$$p_{j_i}(t_1,\ldots,t_k) = t_i + q_i(t_1,\ldots,t_{i-1}) \quad .$$

This says that if $\mu_i(e_j) = \delta_{ij}$, $\mu_i \in V^*$, then if $\lambda_i = \mu_{j_i}$,

$$(*) \quad \lambda_i(\rho(\exp t_1X_1)\cdots\exp(t_kX_k))v) = t_i + q_i(t_1,\ldots,t_{i-1}) \quad .$$

Let $\mathcal{O}_v = G\cdot v$. Then if

$$\Phi(x) = (\lambda_1(x),\ldots,\lambda_k(x)) \quad ,$$

$\Phi\colon \mathcal{O}_v \to \mathbb{R}^k$ is C^∞.

<u>Lemma 3.1</u>. $\Phi\colon \mathcal{O}_v \to \mathbb{R}^k$ is a surjective diffeomorphism.

<u>Proof</u>. If $\Phi(x) = \Phi(x')$, then $\lambda_1(x) = \lambda_1(x')$.

$$x = \rho(\exp t_1X_1 \cdots \exp t_kX_k)v \quad ,$$

$$x' = \rho(\exp t_1'X_1 \cdots \exp t_k'X_k)v \quad ,$$

Thus

$$t_1 = t_1' \quad .$$

Arguing by induction using (*), we find $t_i = t_i'$ for all i. Thus $x = x'$.

(*) also implies that Φ is everywhere regular.

Hence $\Phi(\mathcal{O}_v)$ is open in $\mathbb{R}^k$. We assert that $\Phi(\mathcal{O}_v)$ is closed. Indeed, suppose $v_j \in \Phi(\mathcal{O}_v)$ and

$$\lim_{j\to\infty} v_j = v_0 \in \mathbb{R}^k \quad .$$

$$v_j = \Phi(\rho(\exp t_{i,j}X_1) \cdots \rho(\exp t_{i,j}X_k)v) \quad .$$

Thus

$$\lim_{j\to\infty} t_{i,j} + q_i(t_{i,j},\ldots,t_{i-1,j}) = (v_0)_i \quad .$$

But then

$$\lim_{j\to\infty} t_{i,j} = (v_0)_i \quad .$$

Arguing recursively, we see that each of the $t_{i,j}$ go to a limit in j. Hence, $\Phi(\mathcal{O}_v)$ is closed in $\mathbb{R}^k$. Thus

$$\Phi(\mathcal{O}_v) = \mathbb{R}^k \quad .$$

This completes the proof.

We have proved:

<u>Theorem 3.2</u>. Let G be a connected and simply connected Lie group with Lie algebra g. Let (ρ,V) be a finite dimensional representation of G so that $\rho(X)^k = 0$ for each $X \in g$ for

some k. Let $v \in V$. Let $\mathcal{O}_v = \rho(G)\cdot v$. Then there exists a collection of linearly independent linear forms $\lambda_1,\ldots,\lambda_k$ on V so that if

$$\Phi(x) = (\lambda_1(x),\ldots,\lambda_k(x)) ,$$

then $\Phi\colon \mathcal{O}_v \to \mathbb{R}^k$ is a surjective diffeomorphism. Furthermore, we may extend $\lambda_1,\ldots,\lambda_k$ to a basis $\lambda_1,\ldots,\lambda_n$ of V^* so that $\lambda_{k+1}|\mathcal{O}_v$ is a polynomial in $\lambda_1|\mathcal{O}_v,\ldots,\lambda_k|\mathcal{O}_v$.

<u>Corollary 3.3</u>. In the notation of Theorem 3.2 $\mathcal{O}_v$ is closed in V.

<u>Proof</u>. If $x_j \in \mathcal{O}_v$,

$$\lim_{j\to\infty} x_j = x_0 \in V ,$$

then

$$\lim_{j\to\infty} \lambda_i(x_j) = \lambda_i(x_0) .$$

Thus there is $x_0' \in V$ so that

$$\lim_{j\to\infty} x_j = x_0' .$$

But then there are

$$g_j \in G , \quad \lim_{j\to\infty} g_j \to g_0 , \quad g_0\cdot v = x_0' , \quad x_j = g_j\cdot v .$$

Thus, $x_0' = x_0$. Q.E.D.

Corollary 3.4. Let $v \in V$. Then there is a neighborhood U of v in $\mathbb{R}^n$ so that if $u \in U$, $\dim \mathcal{O}_u \geq \dim \mathcal{O}_v$.

Proof. Let $X_1,\ldots,X_k$, $\lambda_1,\ldots,\lambda_k$ be as above. Define

$$\Phi_u(t_1,\ldots,t_k) = (\lambda_1\rho(\exp t_1X_1\cdots\exp t_kX_k)u,\ldots,$$

$$\lambda_k(\rho(\exp t_1X_1\cdots\exp t_kX_k)u) \quad .$$

Then

$$\det D\Phi_u(0) \neq 0 \quad .$$

Since $u \to \det D\Phi_u(0)$ is continuous, the result follows.

Corollary 3.5. Let

$$d(V) = \max\{\dim \mathcal{O}_v \mid v \in V\} \quad .$$

Then there is a polynomial on V, f, $f \not\equiv 0$ so that if $w \in \{v \in V \mid f(v) \neq 0\}$, $\dim \mathcal{O}_w = d(V)$.

Proof. Fix $v \in V$ so that $\dim \mathcal{O}_v = d(V)$. Set $k = d(V)$. Let $\lambda_1,\ldots,\lambda_k$, $X_1,\ldots,X_k$ be as above, and let Φ_u be as in the proof of Corollary 3.4. Let $f(u) = \det(D\Phi_u(0))$. Then $f(u)$ is clearly a polynomial function on V. If $f(u) \neq 0$, $\dim \mathcal{O}_u \geq k$. Hence, $\dim \mathcal{O}_u = k$. Q.E.D.

4. APPLICATIONS TO THE CO-ADJOINT REPRESENTATION

We now apply the results of Section 3 to the co-adjoint representation (see below) of G a connected, simply connected, Lie group with Lie algebra $\mathfrak{g}$. Recall that if $g \in G$, $f \in \mathfrak{g}^*$ then $g \cdot f = f \circ \mathrm{Ad}\, g^{-1}$. If we set

$$g \cdot f = \mathrm{Ad}^*(g) \cdot f$$

then $(\mathrm{Ad}^*, \mathfrak{g})$ is a unipotent representation of G. $(\mathrm{Ad}^*, \mathfrak{g})$ is the <u>co-adjoint representation</u> of G.

Lemma 2.6 implies that $\exp: \mathfrak{g} \to G$ is a diffeomorphism. We will assume throughout this section that $\exp: \mathfrak{g} \to G$ is the <u>identity mapping</u>. That is, $G = \mathfrak{g}$ with a multiplication and 0 is the multiplication identity.

We fix on $\mathfrak{g}$ a Lebesgue measure, dX. (That is, $\int_{\mathfrak{g}} f(X+Y)\, dX = \int_{\mathfrak{g}} f(X)\, dX$.) On $\mathfrak{g}^*$ we take the dual Lebesgue measure. That is, if

$$\tilde{\varphi}(f) = \int_{\mathfrak{g}} \varphi(X) e^{2\pi i f(X)}\, dX$$

Then

$$\varphi(0) = \int_{\mathfrak{g}^*} \tilde{\varphi}(f)\, df \quad .$$

If $f \in \mathfrak{g}^*$, let

$$G_f = \{g \in G \mid g \cdot f = f\}$$

and

$$\mathfrak{g}_f = \{X \in \mathfrak{g} \mid X \cdot f = 0\}$$

($X \cdot f = 0$ means $f([X,Y]) = 0$ for all $Y \in \mathfrak{g}$.)

Let $f \in \mathfrak{g}^*$ be fixed (for the moment) and let $X_1,\ldots,X_k \in \mathfrak{g}$, $\lambda_1,\ldots,\lambda_k \in \mathfrak{g}^*$ be as in the discussion preceding Lemma 3.1. Let $\Phi\colon \mathcal{O}_f \to \mathbb{R}^k$ be defined by

$$\Phi(X) = (\lambda_1(X),\ldots,\lambda_k(X)) \quad .$$

Lemma 4.1. $\Omega^f = c_f \Phi^*(dy_1 \wedge \cdots \wedge dy_k)$ with c_f a constant depending only on f.

Proof. We first note that if H_1, H_2 are connected subgroups of G and if $\psi\colon H_1 \times H_2 \to G$ $(\psi(h_1,h_2) = h_1 h_2)$ is a surjective diffeomorphism, then there is a constant c so that

$$\int_G f(g)\, dg = \int_{H_1 \times H_2} f(h_1 h_2)\, dh_1\, dh_2$$

(dg, dh_1, dh_2 are some choices of invariant measure on G, H_1, H_2, respectively.) Indeed,

$$\psi^* dg = u(h_1,h_2)\, dh_1 dh_2 \quad .$$

But

$$L_x^* dg = R_x^* dg$$

$$= dg \quad .$$

$(L_x y = xy,\ \ R_x y = yx,\ \ x,y \in G.)$ Hence

$$u(xh_1, h_2 y) = u(h_1, h_2) \quad ,$$

$$\text{for} \quad x \in H_1, \ \ y \in H_2 \quad .$$

Hence u is a constant.

Let

$$G_i = \{\exp t_i X_i \cdots \exp t_k X_k \cdot g \mid t_j \in \mathbb{R},\ j = i, \ldots, k,\ \ g \in G_f\}$$

Then G_i is a connected subgroup of G.

$$G_i = \exp \mathbb{R} X_i \cdot G_{i-1} \quad .$$

Hence,

$$dg_i = c_i dt_i \cdot dg_{i-1} \quad .$$

Applying this recursively, we find

$$dg = c dt_1 \cdots dt_k \, dg_f \quad ,$$

dg_f invariant measure on G_f.

Hence

$$d(gG_f) = c dt_1 \cdots dt_k \quad .$$

But

$$\Phi^*(dy_1 \wedge \cdots \wedge dy_k) = dt_1 \wedge \cdots \wedge dt_k \quad .$$

Q.E.D.

Corollary 4.2. If $f \in g^*$ and if $\varphi \in S(g)$, set

$$\Theta_{O_f}(\varphi) = \int_{O_f} \left(\tilde{\varphi}|_{O_f}\right) \Omega^f \quad .$$

Then $\Theta_{O_f}(\varphi)$ exists for $\varphi \in S(g)$ and Θ_{O_f} defines a tempered distribution on g.

Proof. $S(g) \to S(g^*)$, $\varphi \mapsto \tilde{\varphi}$ is continuous. By the results in Section 3

$$\varphi \mapsto \left(\varphi|_{O_f}\right) \circ \Phi^{-1}$$

is a continuous map of $S(g^*)$ to $S(\mathbb{R}^k)$. Hence Lemma 4.1 implies Corollary 4.2. Q.E.D.

5. THE EXISTENCE OF POLARIZATIONS

In this section we will show that if G is a connected, simply connected Lie group with Lie algebra g and if $f \in g^*$, then there exists a polarization of f (see Definition 1.7). We must first look at some generalities on skew forms.

Let V be a finite dimensional real vector space. Let $B: V \times V \to \mathbb{R}$ be bilinear and skew-symmetric. If W is a subspace of V, set

$$W^{\perp} = \{v \in V \mid B(v,W) = 0\} \quad .$$

Clearly,

$$\dim W + \dim W^{\perp} = \dim V + \dim (V^{\perp} \cap W) .$$

(In fact, let, for $v \in V$, $v^{\#}(u) = B(v,w)$. If $\eta(v) = v^{\#}|_W$ then $\eta: V \to (W/W \cap V^{\perp})^*$, $\ker \eta = W^{\perp}$. Thus, $\dim V = \dim W - \dim W \cap V^{\perp} + \dim W^{\perp}$.)

A subspace $W \subset V$ is said to be <u>isotropic</u> if

$$B|_{W \times W} = 0 ,$$

(i.e., $W \subset W^{\perp}$). If W is maximal among the isotropic subspaces, then $W \supset V^{\perp}$, hence $2 \dim W \leq \dim V + \dim V^{\perp}$.

<u>Lemma 5.1</u>. If $W \subset V$ is an isotropic subspace, then W is maximal isotropic if and only if

$$\dim W = \frac{1}{2} (\dim V + \dim V^{\perp}) .$$

<u>Proof</u>. If $\dim W = \frac{1}{2} (\dim V + \dim V^{\perp})$ then W is maximal isotropic by the above observations. Now B induces a non-degenerate skew form on $V/V^{\perp}$. If $\overline{W}$ is maximal isotropic in $V/V^{\perp}$, then $\dim \overline{W} = \frac{1}{2} \dim(V/V^{\perp})$. Let $\pi: V \to V/V^{\perp}$ be the canonical projection. Set $W = \pi^{-1}\overline{W}$. Then W is isotropic, $V^{\perp} \subset W$ and

$$\begin{aligned} \dim W &= \frac{1}{2} \dim V/V^{\perp} + \dim V^{\perp} \\ &= \frac{1}{2} (\dim V + \dim V^{\perp}) . \end{aligned}$$

Finally, if $U \subset V$ is isotropic containing $V^{\perp}$, then $\pi(U)$ is isotropic. Q.E.D.

Lemma 5.2 (cf. Dixmier [7]). Let $V' \subset V$ be a subspace of codimension one. Let $N' = \{v \in V' |\ B(v, V') = 0\}$.

1) If $V^{\perp} \subset V'$ then $V^{\perp}$ is codimension one in N'. If $W \subset V'$ is maximal isotropic in V', then W is maximal isotropic in V.

2) If $V^{\perp} \not\subset V'$ then $N' = V^{\perp} \cap V'$ and N' has codimension one in $V^{\perp}$. If W is a maximal isotropic subspace in V, then $W \cap V'$ is maximal isotropic in V' and $W \cap V'$ has codimension one in W.

Proof. If $V^{\perp} \subset V'$ then $V^{\perp} \subset N'$. Let $x \in V$ be so that $x \notin V'$. Then

$$V^{\perp} = \{v \in N' |\ B(x, v) = 0\} .$$

Hence $\dim (N'/V^{\perp}) \leq 1$. But $\dim V'/N'$ is even and $\dim V/V^{\perp}$ is even. Hence

$$\dim (N'/V^{\perp}) = 1 .$$

But then

$$\frac{1}{2} (\dim V + \dim V^{\perp}) = \frac{1}{2}((\dim V - 1) + (\dim V^{\perp} + 1))$$

$$= \frac{1}{2} (\dim V' + \dim N') .$$

Hence 1) follows.

2) If $V^{\perp} \not\subset V'$ then there is $x \in V^{\perp}$ so that $\mathbb{R}x + V' = V$. Thus, if $u \in N'$, then

$$B(u,V') = B(u,x) = 0 .$$

Hence

$$N' = V^{\perp} \cap V' .$$

This clearly shows that N' is of codimension one in $V^{\perp}$. Suppose W is maximal isotropic in N. Then $V^{\perp} \subset W$. Hence $x \in W$. If $y \in V'$, $B(y, W \cap V') = 0$, then

$$B(y, W \cap V' + \mathbb{R}x) = 0 .$$

Since $W = W \cap V' + \mathbb{R}x$, $y \in W$. Hence $y \in W \cap V'$. Q.E.D.

<u>Lemma 5.3</u>. Let $\mathfrak{g}$ be a nilpotent Lie algebra. Suppose that the center of $\mathfrak{g}$ is one dimensional. That is,

$$\mathfrak{z} = \{X \in \mathfrak{g} \mid [X,\mathfrak{g}] = 0\}$$

and

$$\dim \mathfrak{z} = 1 .$$

Suppose that $\mathbb{R}Z = \mathfrak{z}$. Then there exist $X,Y \in \mathfrak{g}$ so that $[X,Y] = Z$ and if

$$L = \{W \in \mathfrak{g} \mid [Y,W] = 0\}$$

then L is a codimension one subalgebra of $\mathfrak{g}$.

<u>Proof</u>. Let

$$\mathfrak{g} = \mathfrak{g}_n \supset \mathfrak{g}_{n-1} \supset \cdots \supset \mathfrak{g}_1 \supset (0)$$

be as in Corollary 2.3. Then

$$\mathfrak{z} = \mathfrak{g}_1 \quad .$$

Let $Y \in \mathfrak{g}_2$ be such that

$$\mathfrak{g}_2 = \mathbb{R}Y + \mathbb{R}Z \quad .$$

If $W \in \mathfrak{g}$, then

$$\operatorname{ad} W \cdot Y = f(W)Z \quad ,$$

and $f\colon \mathfrak{g} \to \mathbb{R}$ is linear $([\mathfrak{g},\mathfrak{g}_i] \subset \mathfrak{g}_{i-1})$.

a) $f \not\equiv 0$.

Indeed, if $f = 0$ then $\mathfrak{z} \supset \mathbb{R}Y + \mathbb{R}Z$. But $\dim \mathfrak{g}_2 = 2$, $\dim \mathfrak{z} = 1$.

Let $X \in \mathfrak{g}$ be such that $f(X) = 1$. Then $[X,Y] = Z$. Let

$$L = \{W \in \mathfrak{g} \mid [W,Y] = 0\} \quad .$$

Then

$$L = \operatorname{Ker} f \quad .$$

Hence $\dim \mathfrak{g} = \dim L + 1$. Q.E.D.

Theorem 5.4. If g is a nilpotent Lie algebra and $f \in g^*$, then there exists a polarization of f (see Definition 1.7).

Proof. We prove the result by induction on the dimension of g. If $\dim g = 1$, then g is a polarization of f. Suppose that the theorem has been proved for all nilpotent Lie algebras of dimension $\leq n-1$ and $\dim g = n$. Let z be the center of g $(z = \{X \in g \mid [X,g] = 0\})$.

a) There exists $Y \in z$ so that $f(Y) = 0$, $Y \neq 0$.

Then f defines an element of $(g/\mathbb{R}Y)^*$, $\bar{f}$. Let $\bar{h}$ be a polarization of $\bar{f}$. Then we assert that if $h \supset \mathbb{R}Y$, $h \subset g$, and if $h \to \bar{h}$ under the canonical map, then h is a polarization of f.

Set

$$B_f(u,v) = f([u,v]) \quad ,$$

$$u,v \in g \quad .$$

Clearly, h is isotropic for B_f. If $h \subset V$ with V an isotropic subspace of g relative to B_f, then $V \to \bar{V} \supset \bar{h}$ under the canonical map and $\bar{V}$ is isotropic for $B_{\bar{f}}$. Hence, $\bar{V} = \bar{h}$. But $h \supset \mathbb{R}Y$ hence $V = h$.

b) $f(Z) = 0$, $Z \in z$ implies $Z = 0$.

Then $\dim z = 1$, $z = \mathbb{R}Z$, $f(Z) = 1$. Let $X, Y \in g$ be as in Lemma 5.3.

Let

$$L = \{W \in g \mid [W,Y] = 0\} \quad .$$

Then $\dim L = n-1$. Let $h \subset L$ be a polarization for $f|_L$. Again, set

$$B_f(u,v) = f([u,v]) \quad .$$

Then

$$g^\perp = \{W \in g \mid B_f(W,g) = 0\} \subset L \quad .$$

To see this, we note that $g = \mathbb{R}X \oplus L$, $([X,Y] = Z)$. If $W = cX + v$, $c \in \mathbb{R}$, $v \in L$, and $W \in g^\perp$ then

$$\begin{aligned} B_f(cX+V,\ Y) &= cB_f(X,Y) \\ &= c \quad . \end{aligned}$$

Hence,

$$c = 0 \quad .$$

Thus $L \supset g^\perp$. Now Lemma 5.2 1) implies that h is a polarization for f on g. Q.E.D.

6. UNITARY REPRESENTATIONS OF NILPOTENT LIE GROUPS

Let G be a connected, simply connected nilpotent Lie group with Lie algebra g. Let dX be a Lebesgue measure on g. ($\int_g f(X+Y)\,dX = \int_g f(X)\,dX$.)

Lemma 6.1. Assume that $\exp: g \to G$ is the identity map. Then dX defines a left and right invariant measure on G.

Proof. We prove the result by induction on $\dim G$. The result is clear if $\dim G = 1$. Suppose true for simply connected nilpotent groups of dimension $\leq n - 1$.

Let $z \subset g$ be the center of g. Then z is the center of G. Let $V \subset g$ be a subspace of g so that $g = V \oplus z$. Then if dv is a Lebesgue measure on V and dz is a Lebesgue measure on z, then $dv \wedge dz$ is a Lebesgue measure on G. Let $\psi: G \to G/z$. Then $\psi: V \to G/z$ is a diffeomorphism.

The induction hypothesis implies that if dY is a Lebesgue measure on g/z, then dY is left and right invariant measure on G/z. Clearly

$$\psi^* dY = c\,dv .$$

But

$$dg = c_1 d(gz)dz$$

hence

$$dg = c' dv dz = dX .$$

Q.E.D.

We fix on g a Lebesgue measure dX and look upon dX as left and right invariant measure on G.

We now study the irreducible unitary representations of G. We will use a technique due to Dixmier and Kirillov. Let

$\mathfrak{z}$ be the center of G (and of $\mathfrak{g}$). Let $(\pi, \mathcal{H})$ be an irreducible unitary representation of G. Then

$$\pi(z) = e^{2\pi i\lambda(z)} I$$

with $\lambda: \mathfrak{z} \to \mathbb{R}$ a linear form.

Let

$$\mathfrak{z}_1 = \text{Ker } \lambda \quad .$$

Then if $G_1 = G/\mathfrak{z}_1$ and if $\eta: G \to G/\mathfrak{z}_1$ is the natural map, then $\pi(g) = \pi(g')$ if $\eta(g) = \eta(g')$. Hence, if we define

$$\tilde{\pi}(\eta(g)) = \pi(g) \quad ,$$

then $(\tilde{\pi}, \mathcal{H})$ is an irreducible representation of G_1 and

$$\pi(g) = \tilde{\pi}(\eta(g)) \quad .$$

Hence, if $\mathfrak{z}_1 \neq 0$ we see that $(\pi, \mathcal{H})$ "comes from" a unitary representation of $G/\mathfrak{z}_1$, a simply connected nilpotent Lie group of lower dimension.

Suppose that

$$\mathfrak{z}_1 = 0 \quad .$$

Then $\lambda: \mathfrak{z} \to \mathbb{R}$ is injective. Since $\mathfrak{z} \neq 0$, we see that $\dim \mathfrak{z} = 1$.

Let $\mathfrak{z} = \mathbb{R}Z$. Let $X, Y \in \mathfrak{g}$ be as in Lemma 5.3. Let

$$\mathcal{L} = \{W \in \mathfrak{g} \mid [W, Y] = 0\} \quad .$$

Then $\mathfrak{g} = \mathbb{R}X \oplus \mathcal{L}$ (see Lemma 5.3). Let

$$\mathcal{A} = \mathbb{R}X + \mathbb{R}Y + \mathbb{R}Z$$

and

$$A = \exp \mathcal{A} = \mathcal{A}$$

as a Lie group. Then $\pi|_A$ is a unitary representation of A.

Theorem 4.3.14 implies that there is a Hilbert space $\mathcal{H}_1$ so that $(\pi|_A, \mathcal{H})$ is unitarily equivalent with $(S_\mu, L^2(\mathbb{R}:\mathcal{H}_1))$ with

$$(S_\mu(sX + tY + uZ)\varphi)(z) = \exp(2\pi i\mu(u + t(z-\tfrac{1}{2}s)))\varphi(z - s) \quad ,$$

$\mu = \lambda(Z)$.

We may thus assume that

$$\mathcal{H} = L^2(\mathbb{R}; \mathcal{H}_1)$$

and

$$(\pi(\exp(sX + tY + uZ))\varphi)(z) = \exp(2\pi i\mu(u + t(z-\tfrac{1}{2}s))\varphi(z - s)$$

Let $L = \exp \mathcal{L}$. Then

$$\pi(\ell)\pi(\exp tY) = \pi(\exp tY)\pi(\ell) \quad .$$

Now

$$\pi(\exp tY)f(z) = e^{2\pi i\mu tz} f(z) \quad .$$

Hence if $h \in \mathcal{S}(\mathbb{R})$, then $\pi(\ell)hf = h\pi(\ell)f$. Let for $f \in \mathcal{S}(\mathbb{R})$, $u \in \mathcal{H}_1$,

$$(f \otimes u)(t) = f(t)u \quad .$$

Let

$$g(x) = e^{-x^2/2}$$

$$g \in \mathcal{S}(\mathbb{R}) \quad .$$

Then

$$\pi(\ell)(gf \otimes u) = f\pi(\ell)(g \otimes u)$$

and

$$\pi(\ell)(gf \otimes u) = g\pi(\ell)(g \otimes u) \quad .$$

Hence

$$\pi(\ell)(f \otimes u) = f \cdot g^{-1}\pi(\ell)(g \otimes u) \quad .$$

Here, $g^{-1}\pi(\ell)(g \otimes u)$ is a continuous function from $\mathbb{R}$ to $\mathcal{H}_1$, which is linear in u.

Hence

$$(g^{-1}\pi(\ell)(g \otimes u))(t) = U(\ell,t)u$$

with $U(\ell,t): \mathcal{H}_1 \to \mathcal{H}_1$ linear, $\ell,t \mapsto U(\ell,t)$ strongly continuous. Using the continuity of π we find

1) $$\pi(\ell)f)(t) = U(\ell,t)f(t)$$

for $f \in \mathcal{H}$ and almost all t.

By the definition of $U(\ell,t)$ it is easily seen that $U(\ell,t)$ is unitary.

2) $$U(\ell,\ t-t_0) = U(\exp(-t_0X)\ell\ \exp(t_0X),t) \quad .$$

Indeed,

$$(\pi(\ell)\pi(\exp\ t_0X)f)(t) = U(\ell,t)\cdot f(t-t_0) \quad .$$

Hence,

$$(\pi(\exp\ -t_0X)\pi(\ell)\exp(t_0X)f)(t) = U(\ell,\ t-t_o)f(t) \quad ,$$

hence 2).

2) implies

3) $$U(\ell,t) = U(\exp\ tX\cdot\ell\ \exp(-tX),0) \quad .$$

4) $$U(\ell,0)\cdot U(\ell',0) = U(\ell\ell',0) \quad .$$

Indeed,

$$\begin{aligned}(\pi(\ell\ell')\varphi)(t) &= \pi(\ell)(U(\ell',t)\varphi(t) \\ &= U(\ell,t)U(\ell',t)\varphi(t) \quad .\end{aligned}$$

This says that if $U_0(\ell) = U(\ell,0)$, then $(U_0,\mathcal{H}_1)$ is a unitary representation of L.

We now prove:

5) $$(\pi,\mathcal{H}) \text{ is equivalent with } \operatorname{Ind}_L^G(U_0) \quad .$$

(See Appendix 7.2.)

Let

$$\mathrm{Ind}_L^G (U_0) = (\delta, H_2) .$$

Let $H_{2,0}$ be the space of all $\varphi: G \to H_1$ so that $g \mapsto |\varphi(gL)|^2$ is continuous with compact support.

If $\varphi \in H_{2,0}$, set

$$B\varphi(t) = \varphi(\exp tX) .$$

We note that

$$G = \exp \mathbb{R}X \cdot L$$

and we may assume

$$\int_{G/L} f(gL)\, dgL = \int_{-\infty}^{\infty} f(\exp tXL)\, dt .$$

Thus

$$B: H_{2,0} \to L^2(\mathbb{R}; H_1)$$

is unitary, thus induces a unitary map $H_2 \to L^2(\mathbb{R}; H_1)$.

Clearly,

$$(B\delta(\exp t_0X)\varphi)(t) = \varphi(t - t_0)$$

$$\begin{aligned}(B(\delta(\ell)\varphi))(t) &= \varphi(\ell^{-1} \exp tX) \\ &= \varphi(\exp tX \exp(-tX)\ell^{-1} \exp tX) \\ &= U_0(\exp(-tX)\ell \exp tX)\varphi(\exp tX)\end{aligned}$$

$$= U(\ell,t)B\varphi(t)$$

$$= (\pi(\ell)B\varphi)(t) \quad .$$

Hence $B\circ\delta(g) = \pi(g)\circ B$. Since π is irreducible, B is surjective. This proves 5).

<u>Theorem 6.2</u>. Let G be a connected, simply connected, nilpotent Lie group with Lie algebra $\mathfrak{g}$. Let $(\pi,\mathcal{H})$ be an irreducible unitary representation of G. Then there exists $f \in \mathfrak{g}^*$ and a polarization $\mathfrak{h}$ of f so that if $H = \exp \mathfrak{h}$ and $\chi_f(\exp X) = e^{2\pi i f(X)}$, $X \in \mathfrak{h}$ then $(\pi,\mathcal{H})$ is unitarily equivalent with $\mathrm{Ind}_H^G(\chi_f)$ (see Appendix A.7.2).

<u>Proof</u>. We use the notation of the preceeding discussion. Thus $\pi(z) = e^{2\pi i\lambda(z)}I$, $z \in \mathfrak{z}$. If $\mathfrak{z}_1 = \mathrm{Ker}\,\lambda \neq 0$ then

$$\pi(g) = \tilde{\pi}(\eta(g))$$

with $(\tilde{\pi},\mathcal{H})$ an irreducible unitary representation of $G/\mathfrak{z}_1$.

If there exists

$$\bar{f} \in (\mathfrak{g}/\mathfrak{z}_1)^*$$

and

$$\bar{\mathfrak{h}} \subset \mathfrak{g}/\mathfrak{z}_1$$

a polarization so that $(\tilde{\pi},\mathcal{H})$ is equivalent with $\mathrm{Ind}_{\bar{H}}^{G/\mathfrak{z}}(\chi_{\bar{f}})$ then taking $\mathfrak{h}$ to be the inverse image of $\bar{\mathfrak{h}}$ in $\mathfrak{g}$ under the natural map $\mathfrak{g} \to \mathfrak{g}/\mathfrak{z}_1$ and $f(X) = \bar{f}(X + \mathfrak{z}_1)$ we see that $(\pi, \mathcal{H})$ is unitarily equivalent with $\mathrm{Ind}_H^G(\chi_f)$.

If $\mathfrak{z}_1 = 0$, then we have seen $(\pi, \mathcal{H})$ is unitarily equivalent with $\mathrm{Ind}_L^G(U_0)$. But if there is $\tilde{f} \in \mathcal{L}^*$ and $\tilde{h}$ a polarization of $\tilde{f}$ in $\mathcal{L}$ so that $(U_0, \mathcal{H}_1)$ is equivalent with $\mathrm{Ind}_{\tilde{H}}^L(\chi_{\tilde{f}})$, then $(\pi, \mathcal{H})$ is unitarily equivalent with $\mathrm{Ind}_{\tilde{H}}^G(\chi_{\tilde{f}})$. (See Appendix 7.2.) Let f be any extension of $\tilde{f}$ to $\mathfrak{g}$. Then the proof of Theorem 5.4 implies that h is a polarization of f in $\mathfrak{g}$.

Since the theorem is clearly true for Lie groups of dimension one, our discussion above shows that it is true by induction on the dimension of G. Q.E.D.

<u>Lemma 6.3</u>. Let G be a connected, simply connected nilpotent Lie group with one dimensional center. Let $L \subset G$ be a connected subgroup of codimension one (L is then normal). Then if $Z \subset G$ is the center, $Z \subset L$. Let (σ, V) be an irreducible unitary representation of L so that $\sigma|_Z$ is not trivial. If L has a center of dimension at least two, then $\mathrm{Ind}_L^G(\sigma)$ is irreducible.

<u>Proof</u>. Let

$$\mathrm{Ind}_L^G(\sigma) = (\pi, \mathcal{H}) \quad .$$

Suppose that $A: \mathcal{H} \to \mathcal{H}$ is a bounded operator so that

$$A \circ \pi(x) = \pi(x) \circ A ,$$

$$\text{for } x \in G .$$

We must show that $A = cI$.

Let $\mathcal{g}$ be the Lie algebra of G, $\mathcal{L}$ that of L. Then $[\mathcal{g},\mathcal{L}] \subset \mathcal{L}$. Let

$$\mathcal{L} = \mathcal{L}_{n-1} \supset \mathcal{L}_{n-2} \supset \cdots \supset \mathcal{L}_1 \supset \mathcal{L}_0$$

be such that $[\mathcal{g},\mathcal{L}_i] \subset \mathcal{L}_{i-1}$, $\dim \mathcal{L}_i = i$. Then $[\mathcal{g},\mathcal{L}_1] = 0$. Hence $\mathcal{L}_1 = \mathcal{z}$ the Lie algebra of $\mathcal{z}$. Let $\mathcal{L}_1 = \mathbb{R}Z$. Let $Y \in \mathcal{L}_2$ be such that $\mathbb{R}Y + \mathbb{R}Z = \mathcal{L}_2$. Now

$$\mathcal{g} = \mathbb{R}X \oplus \mathcal{L} ,$$

$$X \in \mathcal{g} .$$

We may assume $[\mathcal{L},Y] = 0$. But then $[X,Y] \neq 0$. Hence we may assume $[X,Y] = Z$.

We realize $\mathrm{Ind}_L^G(\sigma)$ as the space of functions $f\colon \mathbb{R} \to V$ such that

$$\int_{-\infty}^{\infty} |f(s)|^2 \, ds < \infty .$$

and

$$(\pi(\exp tX)f) = f(s - t)$$

and

$$(\pi(\ell)f)(t) = \sigma(\exp(-tX)\ell \exp tX)f(t) ,$$

$$\ell \in L .$$

(This is done by identifying $f(\exp tX)$ with $f(t)$ for $f \in H$.)

Let $S(\mathbb{R}; V)$ denote the space of all Schwartz functions from $\mathbb{R}$ to V. We note that if $\varphi \in S(\mathbb{R})$, $f \in H$, then

$$\int_{-\infty}^{\infty} \varphi(s) f(t-s)\, ds \in S(\mathbb{R}; V) \quad .$$

We will leave it to the reader to show that $A: S(\mathbb{R}; V) \to S(\mathbb{R}; V)$ is continuous relative to the Schwartz topology.

Now

$$A \circ \pi(g) = \pi(g) \circ A \quad .$$

Hence

$$(Af)(t) = (T * f)(t)$$

with $T: S(\mathbb{R}; V) \to V$ a continuous map.

$$(Af)(\exp tX \cdot \ell) = \sigma(\ell)^{-1}(Af)(\exp tX)$$

$$= (A \circ \sigma(\ell)^{-1} f)(\exp tX)$$

implies that we can actually take T to be in $S'(\mathbb{R})$.

Using the fact that

$$\pi(\exp uY) Af = A\pi(\exp uY) f \quad ,$$

$$u \in \mathbb{R} \quad ,$$

we see that

$$T \circ e^{2\pi i u \mu t} = T$$

for $u \in \mathbb{R}$,

$(\sigma(\exp sZ) = e^{2\pi i s\mu}, \ \mu \in \mathbb{R}^X)$. Hence T is supported at 0.
Hence

$$Af = \sum_{k=0}^{m} a_k \frac{d^k}{dt^k} f \quad .$$

This says

$$Af = \sum_{k=0}^{m} (-1)^k a_k \pi(X)^k f \quad .$$

Now,

$$A \circ \pi(Y) = \pi(Y) \circ A \quad .$$

Hence

$$\sum_{k=0}^{m} (-1)^k a_k [\pi(Y), \pi(X)^k] f = 0 \quad .$$

$$[\pi(Y), \pi(X^k)] = \sum_{j=0}^{k-1} \pi(X)^i \pi([Y,X]) \pi(X)^{k-j-1}$$

$$= -2\pi i \mu \sum_{j=0}^{k-1} \pi(X)^{k-1}$$

$$= -2\pi i \mu k \pi(X)^{k-1} \quad .$$

Hence

$$\sum_{k=1}^{m} (-1)^k a_k k\pi(X)^{k-1} f = 0$$

for all $f \in \mathcal{S}(\mathbb{R}; V)$.

This implies $a_k = 0$ for $k > 1$. This proves the lemma.

Lemma 6.4. Let $G, \mathfrak{g}$ be as usual. Let $f \in \mathfrak{g}^*$. Then there exists a polarization of $f, \mathfrak{h}$ in $\mathfrak{g}$ so that $\mathrm{Ind}_H^G(\chi_f)$ is irreducible.

Proof. By induction on $\dim G$. As before the only case that must be analyzed is $\dim \mathfrak{z} = 1$ and $f(\mathfrak{z}) \neq 0$. Let

$$A = \mathbb{R}X + \mathbb{R}Y + \mathbb{R}Z ,$$

$$[X,Y] = Z$$

be as usual.

$$L = \{W \in \mathfrak{g} \mid [W,Y] = 0\}$$

L codimension 1. There is a polarization of $f|_L$, $\mathfrak{h} \subset \mathfrak{L}$, so that $\mathrm{Ind}_H^L(\chi_f)$ is irreducible by the inductive hypothesis. Arguing as in the proof of Theorem 5.4, $\mathfrak{h}$ is a polarization of f in $\mathfrak{g}$. Lemma A.7.II.2 implies

$$\mathrm{Ind}_L^G(\mathrm{Ind}_H^L(\chi_f)) = \mathrm{Ind}_H^G(\chi_f)$$

is irreducible. Q.E.D.

Lemma 6.5. Let $G, \mathfrak{g}$ be as usual. If $f \in \mathfrak{g}^*$ and $\mathfrak{h}$ is a polarization of f, set $\pi_{\mathfrak{g}}(f,\mathfrak{h}) = \mathrm{Ind}_H^G(\chi_f)$. Then, if $\mathfrak{h}_1, \mathfrak{h}_2$ are polarizations of f, $\pi_{\mathfrak{g}}(f,\mathfrak{h}_1) \equiv \pi_{\mathfrak{g}}(f,\mathfrak{h}_2)$.

Proof. By induction on dimension $\mathfrak{g}$. As usual we come to the case $\dim \mathfrak{z} = 1$ and $f(\mathfrak{z}) \neq 0$. Let

$$A = \mathbb{R}X + \mathbb{R}Y + \mathbb{R}Z$$

as usual,

$$L = \{W \in \mathfrak{g} \mid [W,Y] = 0\}$$

and

$$\dim L = \dim \mathfrak{g} - 1 \quad .$$

We may assume

$$f(X) = f(Y) = 0 \quad .$$

Let $\mathfrak{h}$ be a polarization of f. If $W \in \mathfrak{g}$,

$$\mathrm{ad}\, W \cdot Y = \lambda(W) Z \quad ,$$

$$\lambda \in \mathfrak{f}^* \quad .$$

$\mathrm{Ker}\, \lambda = L$. Thus

$$\mathfrak{h} \cap L = \mathrm{Ker}\, \lambda|_{\mathfrak{h}} \quad .$$

Hence

$$\dim(\mathfrak{h} \cap L) \geq \dim \mathfrak{h} - 1 \quad .$$

a) $h \cap L = h$ for all polarizations of f. Then

$$\mathrm{Ind}_H^G(\chi_f) = \mathrm{Ind}_L^G(\pi_L(f,h))$$

which is independent of h.

b) $\dim h \cap L = \dim h - 1$. Set $h' = h \cap L$. Then we assert that $h_1 = \mathbb{R}Y + h'$ is a polarization of $f|_L$. Indeed, if $a,b \in \mathbb{R}$, $u,v \in h'$ then

$$f([aY+u,\ bY+v]) = f([u,v]) = 0 \quad .$$

Hence $f([h_1,h_1]) = 0$. Suppose that $Y \in h'$. If $W \in h$, $W \notin L$, then $W = aX + W_1$, $W_1 \in L$, $a \neq 0$, $f([W,Y]) = af([X,Y]) = af(Z) \neq 0$. Thus $Y \notin h'$. Hence $\dim h_1 = \dim h$. Thus h_1 is a polarization.

We assert that

$$\pi_g(f,h) \equiv \pi_g(f,h_1) \quad .$$

Once this is shown, then the result follows as in a). We may assume that $X \in h$. (There clearly is $X_1 \in h$ so that $\lambda(X_1) = 1$, replace X by X_1.)

Hence $h = \mathbb{R}X + h'$, $h_1 = \mathbb{R}Y + h'$, $[X,Y] = Z$, $\mathbb{R}Z = z$. Let

$$k = h + \mathbb{R}Y \quad .$$

Then k is a subalgebra of g (indeed, $\mathrm{ad}\, Y \cdot h' \subset h'$, $[Y,X] = Z \in h'$). Let K be the connected subgroup of G corresponding to k. We must show $\mathrm{Ind}_H^K(\chi_f) \equiv \mathrm{Ind}_{H_1}^K(\chi_f)$. Since then

$$\mathrm{Ind}_{H_1}^G(\chi_f) \equiv \mathrm{Ind}_K^G(\mathrm{Ind}_{H_1}^K(\chi_f))$$

$$\equiv \mathrm{Ind}_K^G(\mathrm{Ind}_H^K(\chi_f))$$

$$\equiv \mathrm{Ind}_H^G(\chi_f) \quad .$$

But

$$\mathrm{Ind}_H^K(\chi_f) = (\pi, \mathcal{H}) \quad ,$$

$$\mathrm{Ind}_{H_1}^K(\chi_f) = (\pi_1, \mathcal{H}_1)$$

with $\mathcal{H}$ the space of all $\varphi\colon K \to \mathbb{C}$

$$\varphi(kh) = \chi_f(h)^{-1}\varphi(k) \quad ,$$

$$\int_{K/H} (\varphi(k))^2 \, d(kH) < \infty$$

$\mathcal{H}_1$ the space of all $\varphi\colon K \to \mathbb{C}$,

$$\varphi(kh_1) = \chi_f(h_1)^{-1}\varphi(k)$$

and

$$\int_{K/H_1} |\varphi(k)|^2 \, d(kH_1) < \infty$$

We write H as $\exp tX\cdot H'$, H_1 as $\exp tY\cdot H'$. $K = \exp tX\cdot\exp sY\cdot H'$. Hence $\varphi \in \mathcal{H}$ satisfies

$$\varphi(\exp tX \exp sY\cdot h') = \varphi(\exp s(Y+stZ))\chi_f(\exp tX\cdot h')$$

$$= e^{-2\pi istf(Z)}\chi_f(h')^{-1}\varphi(\exp sY)$$

and

$$\int_{-\infty}^{\infty} |\varphi(\exp sY)|^2\, ds < \infty$$

$\varphi \in \mathcal{H}_1$ satisfies

$$\varphi(\exp tX \exp sY\cdot h') = \chi_f(h')^{-1}\varphi(\exp tX)$$

and

$$\int_{-\infty}^{\infty} |\ (\exp tX)|^2\, dt$$

Define for $\varphi \in \mathcal{H}$,

$$(A\varphi)(\exp tX) = \int_{-\infty}^{\infty}\varphi(\exp sY)\, e^{-2\pi istf(Z)}\, ds \quad .$$

Then extend

$$(A\varphi)(\exp tX \exp sYh') = \chi_f(h')^{-1}A\varphi(\exp tX) \quad .$$

We have $A\varphi \in \mathcal{H}_1$. Also,

$$\pi_1(h')(A\varphi)(\exp tX) = (A\varphi)(h'^{-1}\exp tX)$$

$$= \quad (A\varphi)(\exp tX \exp(-tX)h'^{-1} \exp tX)$$

$$= \quad \chi_f(h')(A\varphi)(\exp tX) \quad .$$

$$(A(\pi(h')\varphi))(\exp tX) \quad = \quad \int_{-\infty}^{\infty} (\pi(h')\varphi)(\exp sY)e^{2\pi istf(Z)} \, ds$$

$$= \quad \int_{-\infty}^{\infty} \varphi(\exp sYh'^{-1})e^{-2\pi istf(Z)} \, ds$$

$$= \quad \chi_f(h') \int_{-\infty}^{\infty} \varphi(\exp sY)e^{2\pi istf(Z)} \, ds \quad .$$

Thus

$$A\circ\pi(h') \quad = \quad \pi_1(h')\circ A \quad .$$

$$(A\circ\pi(\exp uY)\varphi)(\exp tX) \quad = \quad \int_{-\infty}^{\infty} (\pi(\exp uY)\varphi)(\exp sY)e^{-2\pi istf(Z)} \, ds$$

$$= \quad \int_{-\infty}^{\infty} \varphi(\exp(s-u)Y)e^{-2\pi istf(Z)} \, ds$$

$$= \quad \int_{-\infty}^{\infty} \varphi(\exp sY)e^{-2\pi i(s+u)tf(Z)} \, ds$$

$$= \quad e^{-2\pi iutf(Z)} \int_{-\infty}^{\infty} \varphi(\exp sY)e^{-2\pi istf(Z)} \, ds$$

$$= \quad e^{-2\pi iutf(Z)} (A\varphi)(\exp tX) \quad .$$

$$\pi_1(\exp uY)(A\varphi)(\exp tX) = (A\varphi)(\exp(-uY)\exp tX)$$

$$= (A\varphi)(\exp tX \exp utZ)$$

$$= e^{-2\pi iutf(Z)} A\varphi(\exp tX) .$$

Hence

$$A\circ\pi(\exp uY) = \pi_1(\exp uY)\circ A .$$

Finally,

$$(A\circ\pi(\exp uX)\varphi)(\exp tX) = \int_{-\infty}^{\infty} \pi(\exp uX)\varphi(\exp sY)e^{-2\pi istf(Z)}ds$$

$$= \int_{-\infty}^{\infty} \varphi(\exp(-uX)\exp sY)e^{-2\pi istf(Z)}ds$$

$$= \int_{-\infty}^{\infty} \varphi(\exp sY \exp(-usZ)e^{-2\pi istf(Z)}dx$$

$$= \int_{-\infty}^{\infty} \varphi(\exp sY)e^{-2\pi is(t-u)f(Z)}\, ds$$

$$= (A\varphi)(\exp(t-u)X) .$$

Hence

$$A\circ\pi(\exp uX) = \pi(\exp uX)\circ A .$$

This completes the proof of the lemma.

<u>Lemma 6.6</u>. Let $G, \mathcal{g}$ be as usual. If $f_1, f_2 \in \mathcal{g}^*$, $g \cdot f_1 = f_2$, $g \in G$ and h_1, h_2 are polarizations of f_1, f_2 respectively, then $\pi(f_1, h_1) \equiv \pi(f_2, h_2)$.

<u>Proof</u>. $f_2(X) = f_1(\mathrm{Ad}(g^{-1})X)$, $X \in \mathcal{g}$. Define

$$A\colon \pi(f_1, h_1) \to \pi(f_2, \mathrm{Ad}(g)h_1)$$

by

$$(A\varphi)(X) = \varphi(Xg) \quad .$$

Then

$$(A\varphi)(Xgh_1g^{-1}) = \varphi(Xgh_1)$$

$$= e^{-2\pi i f_1(\log h_1)} (A\varphi)(X) \quad .$$

Let $h_1' = \mathrm{Ad}(g)h_1$. Then

$$(A\varphi)(X \exp h) = e^{-2\pi i f_2(h)} (A\varphi)(X)$$

$$\text{for } h \in h_1' \quad .$$

A defines the equivalence. Now, $\mathrm{Ad}(g)h_1$ is a polarization of f_2 (this is clear since if A is a polarization of f_2, $\mathrm{Ad}(g)^{-1}A$ is maximal isotropic for $B_{f_1}(X,Y) = f_1([X,Y])$). Lemma 6.5 implies that $\pi(f_2, \mathrm{Ad}(g)h_1) \equiv \pi(f_2, h_2)$. Q.E.D.

Our last task in the classification of irreducible unitary representations of G is to compute the characters. We will do this in the next section.

7. THE CHARACTER FORMULA

In the last section we established a one-to-one correspondence between the orbits of G acting on $\mathfrak{g}^*, \mathcal{O}$ and the equivalence classes of irreducible unitary representations of G. That is, if $\mathcal{O}$ is an orbit of G acting on $\mathfrak{g}^*$, set $\pi(\mathcal{O})$ equal to the equivalence class of $\pi(f,\mathfrak{h})$ with $f \in \mathcal{O}$ and $\mathfrak{h}$ a polarization of f. In this section we show that if $\omega \in \hat{G}$ (the set of equivalence classes of irreducible representations of G), then ω is trace class and we compute the character of $\omega = \pi(\mathcal{O})$.

<u>Theorem 7.1</u>. If $\mathcal{O} \subset \mathfrak{g}^*$ is an orbit, then $\pi(\mathcal{O})$ is of trace class and

$$\Theta_{\pi(\mathcal{O})}(f) = \int_{\mathcal{O}} (\tilde{f}|_{\mathcal{O}})\ \Omega^{\varphi}$$

$$\varphi \in \mathcal{O} \quad \text{and} \quad f \in C_c^{\infty}(G) \quad .$$

(Here we take $G = \mathfrak{g}$ with $\exp: \mathfrak{g} \to G$ the identity map, $dg = dX$ and df on $\mathfrak{g}^*$ the dual measure:

$$\int_{\mathfrak{g}^*} \int_{\mathfrak{g}} \varphi(X) e^{-2\pi i f(X)}\, dX\, df = \varphi(0) \quad .)$$

<u>Proof</u>. By induction on $\dim \mathfrak{g}$. We actually prove a stronger result. We show that if $\pi \in \pi(\mathcal{O})$, then $\pi(\varphi)$ is trace class if $\varphi \in \mathcal{S}(\mathfrak{g})$. We also show in the induction that

if H is the Hilbert space on which π acts there is a basis of H, $\{v_j\}$ so that if $\varphi \in S(g)$ then

$$\sum_{ij} |\langle\pi(\varphi)v_i, v_j\rangle| < \infty$$

If $\dim g = 1$ this result is obvious. Suppose that the result has been proved for $1 \leq \dim g < r$. Suppose that $\dim g = r$.

Let $f \in \mathcal{O}$ and $\pi = \pi(f,h)$ with h a polarization of f. Let z be the center of g. Let $z_1 = \operatorname{Ker} f|_z$.

a) $\dim z_1 \geq 1$.

Set $Z_1 = \exp z_1$. Let $\eta\colon G \to G/Z_1 = G_1$, $\eta\colon g \to g/z_1 = g$ be the natural maps.

Define

$$\tilde{\pi}(gZ_1) = \pi(g) ,$$

$$g \in G .$$

Then if $f_1(X+z_1) = f(X)$, $h_1 = \eta(h)$, $\tilde{\pi} = \pi(f_1,h_1)$. The inductive hypothesis implies that if $\varphi \in S(g_1)$, then $\tilde{\pi}(\varphi)$ is trace class and

$$\operatorname{tr} \tilde{\pi}(\varphi) = \int_{\mathcal{O}_{f_1}} \left(\hat{\varphi}\Big|_{\mathcal{O}_{f_1}}\right)\Omega^{f_1} .$$

Let $\varphi \in S(g)$. Then

$$\pi(\varphi) = \int_{\mathfrak{g}} \varphi(X)\pi(\exp X)\, dX$$

$$= \int_{\mathfrak{g}/\mathfrak{z}_1} \left(\int_{\mathfrak{z}_1} \varphi(X+z)\, dz\right) \tilde{\pi}(\exp(X+\mathfrak{z}_1))\, d(X+\mathfrak{z}_1)$$

$$= \tilde{\pi}(\tilde{\varphi})$$

with

$$\tilde{\varphi}(\eta(X)) = \int_{\mathfrak{z}_1} \varphi(X+z)\, dx \quad .$$

Now, if $G_f = \{g \in G \mid g\cdot f = f\}$, $G_f \supset Z_1$. Hence

$$G/G_f = G_1/(G_1)_{f_1} \quad ,$$

and it is clear that

$$\Omega^{f_1} = \Omega^f$$

on $G/G_f = G_1/(G_1)_{f_1}$. Now,

$$\hat{\varphi}(x) = \int_{\mathfrak{g}/\mathfrak{z}_1} \tilde{\varphi}(\overline{X}) e^{2\pi i x(\overline{X})}\, d\overline{X}$$

$$= \int_{\mathfrak{g}} \varphi(X) e^{2\pi i x(\eta(X))}\, dX \quad ,$$

for $x \in \mathfrak{g}_1^*$.

If $x \in \mathcal{O}^f$, then $x = x' \circ \eta$, $x' \in \mathcal{O}^{f_1}$. Hence,

$$\operatorname{tr} \tilde{\pi}(\tilde{\varphi}) = \int_{\mathcal{O}^f} \left(\hat{\varphi}\Big|_{\mathcal{O}^f}\right)\Omega^f .$$

This proves the result in this case.

b) $f: z \to \mathbb{R}$ is bijective.

Then we can find X, Y, Z, $\mathbb{R}Z = z$, $f(Y) = 0$, $L = \{W \in g \mid [\overline{W}, Y] = 0\}$ and $\dim L = \dim g = -1$. $\mathbb{R}X + L = g$. Also there is $h \subset L$ a polarization for f. Then

$$\begin{aligned} \pi &= \pi(f,h) \\ &= \operatorname{Ind}_L^G (\pi_L(f_1,h)) , \end{aligned}$$

$$f_1 = f|_L .$$

Let $\varphi \in S(g)$. Then we use

$$d(tX + W) = c\,dt\,dW ,$$

$$t \in \mathbb{R}, \quad W \in g .$$

c is the scalar that makes dtdW agree with the <u>fixed</u> dU we have chosen on g. By renormalizing X and Y we may take $c = 1$. Let $(\pi_L, H_L) = \pi_L(f_1, h)$. If $\varphi \in S(g)$, then

$$(\pi(\varphi)h)(\exp tX) = \int_G \varphi(u)h(u^{-1} \exp tX)\, du$$

$$= \int_G \varphi(\exp tXu^{-1})h(u)\, du$$

$$= \int_{-\infty}^{\infty} \int_L \varphi(\exp tX\ell^{-1} \exp(-sX))\pi_L(\ell)^{-1}h(\exp sX)\, ds\, d\ell$$

$$= \int_{-\infty}^{\infty} \int_L \varphi(\exp tX\ell \exp(-sX))\pi_L(\ell)h(\exp sX)\, ds\, d\ell \quad .$$

Set

$$K_\varphi(t,s) = \int_L \varphi(\exp tX\ell \exp(-sX))\pi_L(\ell)\, d\ell \quad .$$

Then $K_\varphi(t,s)$ is, by the inductive hypothesis, of trace class for all $t,s \in \mathbb{R}$ and

$$(\pi(\varphi)h)(\exp tX) = \int_{-\infty}^{\infty} K\varphi(t,s)h(\exp sX)\, ds \quad .$$

Set

$$\varphi(t:\ell:s) = \varphi(\exp tX\ell \exp(-sX)) \quad .$$

If h_m denotes the m-th Hermite function, then set

$$\varphi_{m,n}(\ell) = \int_{-\infty}^{\infty} \varphi(t,\ell,s)h_m(t)h_n(s)\, dt\, ds$$

Then $\varphi_{k,m} \in S(L)$ and

$$\sum_{m,n=0} h_n(t)\varphi_{m,n}(\ell)h_n(s) = \varphi(t:\ell:s) \quad .$$

Now,

$$K_\varphi(t,s) = \sum_{m,n=0}^{\infty} h_m(t)\pi_L(\varphi_{m,n})h_n(s)$$

with convergence in the operator topology. Let $\{v_j\}$ be a fixed orthonormal basis of H_L so that if $\varphi \varepsilon S(L)$ then

$$\sum_{r,s} |<\pi_L(\varphi)v_i,v_j>| < \infty$$

Set

$$w_{i,j} = h_i v_j \quad .$$

We compute

$$<\pi(\varphi)h_i v_j, h_k v_p> = <\pi_L(\varphi_{ik})v_j,v_p>$$

Using the results in the Appendix to Chapter 5 we see that

$$\sum |<\pi(\varphi)h_i v_j,h_k v_p>| < \infty$$

We can now compute the trace

$$\sum <\pi(\varphi)h_i v_j,h_i v_j> = \sum_j \int_{-\infty}^{\infty} <K_\varphi(t,t)v_j,v_j>$$

$$= \int_{-\infty}^{\infty} \operatorname{tr} K_\varphi(t,t)\, dt \quad .$$

To complete the induction we must compute $\operatorname{tr} \pi(\varphi)$, $\varphi \in S(g)$. Let $f_1 = f|_L$. Let

$$L_f = \{\ell \in L \mid \ell \cdot f = f\} ,$$

$$L_{f_1} = \{\ell \in L \mid \ell \cdot f_1 = f_1\} .$$

I) $$L_{f_1} = L_f \cdot \{\exp tY \mid t \in \mathbb{R}\} , \quad L_f = G_f .$$

This follows from the fact that $L_f \subset L_{f_1}$ and $Y \cdot f_1 = 0$, $Y \cdot f \neq 0$. $\dim g_f = \dim L_{f_1} - 1$. (See Lemma 5.2.)

II) There is a set of linearly independent vectors $X_1, \ldots, X_r \in L$ so that if $g \in G$,

$$g = \exp tX \exp t_1X_1 \cdots \exp t_rX_r \exp sY \cdot \ell_f ,$$

$$\ell_f \in L_f$$

and the map $\mathbb{R}^{r+2} \times L_f \to G$, $t, t_1, \ldots, t_r, s, \ell_f \to \exp tX \exp t_1X_1 \cdots \exp t_rX_r \exp sX \cdot \ell_f$ is a diffeormorphism on $G/G_f = G/L_f$, Ω^f goes to $\frac{1}{\mu}$ dtds $dt_1 \cdots dt_r$, $\mu = f(Z)$.

We now compute

$$\Theta_{0^f}(\varphi) = |\mu| \int \tilde{\varphi}(\exp tX \exp t_1X_1 \cdots \exp t_rX_r \exp sY \cdot f) \, dt \, ds \, dt_1 \cdots dt_r$$

$$= \quad |\mu| \int \tilde{\varphi}(\exp\, tX \exp\, sY \exp\, t_1X_1 \cdots \exp\, t_rX_r \cdot f)\; dt\, ds\, dt_1 \cdots dt_r$$

$(\omega^f(X^*,Y^*) = \mu = f(Z))$.

Set

$$\ell(t_1,\ldots,t_r) = \ell = \exp\, t_1X_1 \cdots \exp\, t_rX_r \quad .$$

Then if $W \in g$, $W = aX + W_1$, $W_1 \in L$

$$\begin{aligned}(\exp\, tX \exp\, sY\cdot\ell\cdot f)(W) &= (\ell\cdot f)(e^{-s\,\mathrm{ad}\,Y}\, e^{-t\,\mathrm{ad}\,X}\, W) \\ &= (\ell\cdot f)(e^{-s\,\mathrm{ad}\,Y} aX + e^{-t\,\mathrm{ad}\,X}\, W_1) \\ &= (\ell\cdot f)(aX + asZ + e^{-t\,\mathrm{ad}\,X}\, W_1) \\ &= (\ell\cdot f)(aX) + (\ell\cdot f)(e^{-t\,\mathrm{ad}\,X}\, W_1) \\ &\qquad + as\mu\end{aligned}$$

$((\ell\cdot f)(Z) = f(Z) = \mu.)$

Hence

$$\begin{aligned}\Theta_{\mathcal{O}_f}(\varphi) &= |\mu| \int_{-\infty}^{\infty} \int_L \varphi(aX+W_1)e^{2\pi i(\exp\, tX \exp\, sY\cdot\ell\cdot f)(aX+W_1)} \\ &\qquad \cdot da\, dW_1\, dt\, ds\, dt_1 \cdots dt_r \\ &= |\mu| \iint \varphi(aX+W_1)e^{2\pi i((\ell\cdot f)(aX)+as\mu)}\; e^{2\pi i(\ell\cdot f)(e^{-t\,\mathrm{ad}\,X}W_1)} \\ &\qquad \cdot da\, dW_1\, dt\, ds\, dt_1 \cdots dt_r \quad .\end{aligned}$$

Now if $\psi \in \mathcal{S}(\mathbb{R})$,

$$\int_{-\infty}^{\infty}\int_{-\infty}^{\infty} \psi(a)e^{2\pi i((\ell\cdot f)(aX) + as\mu)}\, da\, ds$$

$$= \int_{-\infty}^{\infty}\int_{-\infty}^{\infty} \psi(a)e^{2\pi ia((\ell\cdot f)(X) + s\mu)}\, da\, ds$$

$$= \int_{-\infty}^{\infty}\int_{-\infty}^{\infty} \psi(a)e^{2\pi ias\mu}\, da\, ds \quad , \qquad (\mu \neq 0)$$

$$= \frac{1}{|\mu|}\psi(0) \quad .$$

Thus

$$\Theta_{\mathcal{O}^f}(\varphi) = \int_{-\infty}^{\infty}\int_{\mathcal{O}_L^{f_1}}\int_{L} \varphi(W_1)e^{2\pi i(\ell\cdot f)(e^{-t\,\mathrm{ad}\,X}\cdot W_1)}\, dW_1\, d\Omega^{f_1}\, dt$$

$$= \Theta_{\mathcal{O}_L^{f_1}}(\tilde{\varphi}_t) \quad .$$

$$\tilde{\varphi}_t(\ell) = \varphi(\exp tX\cdot\ell\, \exp -tX) \quad .$$

But $\operatorname{tr} K_\varphi(t,t) = \Theta_{\mathcal{O}_L^{f_1}}(\tilde{\varphi}_t)$. Hence

$$\int_{-\infty}^{\infty} \operatorname{tr} K_\varphi(t,t)\, dt = \Theta_{\mathcal{O}^f}(\varphi) \quad .$$

And we have $\Theta_{\pi(f,h)}(\varphi) = \Theta_{\mathcal{O}^f}(\varphi)$. Q.E.D.

We can now state the full Kirillov theorem (which we have proved).

<u>Theorem 7.2</u>. Let G be a connected, simply connected nilpotent Lie group with Lie algebra $\mathfrak{g}$. We assume $\exp\colon \mathfrak{g} \to G$ is the identity. If $f \in \mathfrak{g}^*$ then there exists $\mathfrak{h}$ a polarization of $f,\mathfrak{h}$ and

1) $\pi(f,\mathfrak{h})$ is irreducible. $\pi(f,\mathfrak{h}) \equiv \pi(f',\mathfrak{h}')$ if $f,f' \in \mathcal{O}_f$, $\mathfrak{h}'$ a polarization of f'. Set $[\pi(f,\mathfrak{h})]$ (the equivalence class in $\hat{G}$) equal to $\pi(\mathcal{O}_f)$. $\pi(\mathcal{O})$ is trace class for each $\mathcal{O} \subset \mathfrak{g}^*$ a G-orbit.

2) $\Theta_{\pi(\mathcal{O}_f)} = \Theta_{\mathcal{O}_f}$. Hence $\pi(\mathcal{O}) = \pi(\mathcal{O}')$ if and only if $\mathcal{O} = \mathcal{O}'$.

3) If $\omega \in \hat{G}$ then $\omega = \pi(\mathcal{O})$ for some orbit $\mathcal{O} \subset \mathfrak{g}^*$.

APPENDIX I TO CHAPTER 7

TRACE CLASS OPERATORS

Let $(H,\langle\ ,\ \rangle)$ be a separable Hilbert space (i.e., it has a countable orthonormal basis). A bounded operator $A: H \to H$ is said to be of trace class if

1) If $\{\varphi_n\}$ is an orthonormal basis of H, $\Sigma|\langle A\varphi_n,\varphi_n\rangle| < \infty$

2) $\Sigma\langle A\varphi_n,\varphi_n\rangle = \mathrm{tr}\, A$ is independent of the orthonormal basis $\{\varphi_n\}$.

An operator $A: H \to H$ is said to be summable if there exists $\{\varphi_n\}$ an orthonormal basis of A so that

$$\sum_{n,m} |\langle A\varphi_n,\varphi_m\rangle| < \infty$$

Lemma A.7.I.1. If A is summable, then A is trace class. Furthermore, if B,C are bounded operators on H, then BAC, ABC, BCA are all trace class and $\mathrm{tr}\, BAC = \mathrm{tr}\, ACB = \mathrm{tr}\, CBA$.

Proof. Let $\{\varphi_n\}$ be an orthonormal basis of H so that $\Sigma_{n,m} |\langle A\varphi_n,\varphi_m\rangle| < \infty$. Let

$$b_{ij} = \langle B\varphi_i,\varphi_j\rangle \quad , \qquad c_{ij} = \langle C\varphi_i,\varphi_j\rangle \quad .$$

Then

$$\sum_{i,j,n \geq N} b_{in}c_{nj}\langle A\varphi_i,\varphi_j\rangle \leq \sum_{i,j=1}^{N}\left(\sum_{n=1}^{N}|a_{in}|^2\right)^{1/2}\left(\sum_{n=1}^{N}|c_{jn}|^2\right)^{1/2}|\langle A\varphi_i,\varphi_j\rangle| .$$

Now

$$\sum_{n=1}^{N}|a_{in}|^2 \leq ||B\varphi_i||^2 \leq ||B||^2$$

$$\sum_{n=1}^{N}|c_{jn}|^2 \leq ||C^*\varphi_j||^2 \leq ||C^*||^2 .$$

Hence

$$\sum_{i,j,n=1}^{N}|b_{ij}c_{nj}\langle A\varphi_i,\varphi_j\rangle| \leq ||B||\,||C|| \sum_{i,j=1}^{N}|\langle A\varphi_i,\varphi_j\rangle| .$$

This implies that

$$(*) \qquad \sum_{i,j,n=1}^{\infty} b_{in}c_{nj}\langle A\varphi_i,\varphi_j\rangle$$

is absolutely convergent.

Now $\Sigma\langle BAC\varphi_n,\varphi_n\rangle$, $\Sigma\langle CBZ\varphi_n,\varphi_n\rangle$ and $\Sigma\langle ACB\varphi_n,\varphi_n\rangle$ are all rearrangements of (*). If $U: H \to H$ is a unitary bijection, then

$$\sum_{n=1}^{\infty} \langle BACU\varphi_n, U\varphi_n \rangle = \sum_{n=1}^{\infty} \langle U^{-1}BACU\varphi_n, \varphi_n \rangle$$

$$= \sum_{n=1}^{\infty} \langle CUU^{-1}BA\varphi_n, \varphi_n \rangle$$

$$= \sum_{n=1}^{\infty} \langle CVA\varphi_n, \varphi_n \rangle$$

$$= \sum_{n=1}^{\infty} \langle BAC\varphi_n, \varphi_n \rangle$$

Hence BAC is trace class. Similarly for CBA and ACB. Q.E.D.

Recall that a bounded linear operator $A: H \to H$ is said to be compact if, whenever $\{\varphi_j\}$ is a sequence so that $\|\varphi_j\| \le C < \infty$, for all j, then there is a convergent subsequence of $\{A\varphi_j\}$. We also recall (cf. [26]) :

Lemma A.7.I.2. If $A: H \to H$ is compact and $A^* = A$, then there is an orthonormal basis φ_n of H and a sequence of elements $\lambda_n \in \mathbb{R}$ so that

1) $A\varphi_n = \lambda_n \varphi_n$
2) If j is fixed so that $\lambda_j \neq 0$, there are only a finite number of n so that $\lambda_n = \lambda_j$.
3) If $\dim H = \infty$, $\lim_{j\to\infty} \lambda_{n_j} = 0$ where $\{\lambda_{n_j}\}$ is the subsequence of the nonzero λ_n.

Lemma A.7.I.3. If A is trace class, then A is compact. If A is compact and $A^* = A$, then A is trace class if and only if A is summable.

Proof. If A is trace class then it is easy to see that A is compact. Suppose that $A^* = A$ and A is trace class. We must show that A is summable. Let λ_n, φ_n be as in Lemma A.7.I.2. Then

$$|\langle A\varphi_n, \varphi_m\rangle| = |\lambda_n| \delta_{n,m}$$

Thus

$$\sum_{n,m} |\langle A\varphi_n, \varphi_m\rangle| = \sum_n |\langle A\varphi_n, \varphi_n\rangle|$$

The converse is a consequence of Lemma A.7.I.1.

Lemma A.7.I.4. Let $A: H \to H$ be a compact linear operator. Then there exists $U: H \to H$ a unitary bijection and $T: H \to H$ a compact operator so that $T^* = T$ and $\langle T\varphi, \varphi\rangle \geq 0$ for $\varphi \in H$ and $A = UT$.

Proof. Let φ_n be an orthonormal basis for H and $\lambda_n \geq 0$ so that $A^*A\varphi_n = \lambda_n \varphi_n$. Define

$$T\varphi_n = \lambda_n^{1/2} \varphi_n \quad .$$

Then T defines a compact linear operator on H satisfying $T^* = T$, $\langle T\varphi, \varphi\rangle \geq 0$. We note that $T^2 = A^*A$. Hence

$$||A\varphi||^2 = ||T\varphi||^2 \quad .$$

Thus $T\varphi = 0$ if and only if $A\varphi = 0$. Define

$$Ux = x$$

on Ker T. Let

$$UTx = Ax$$

$$\text{for } x \in (\text{Ker } T)^{\perp} \quad .$$

Then U is well defined $((\text{Ker } T)^{\perp} = \text{Im } T)$. If $\lambda_n = 0$, then set $U\varphi_n = \varphi_n$. If $\lambda_n \neq 0$ then

$$U\varphi_n = \frac{1}{\lambda_n^{1/2}} A\varphi_n \quad .$$

It is easily seen that

$$<U\varphi_j, U\varphi_k> = \delta_{jk} \quad .$$

Hence U is unitary. Q.E.D.

<u>Lemma A.7.I.5</u>. Let $A: H \to H$ be of trace class. Then if $A = UT$ as in Lemma A.7.I.4, T is trace class, hence summable (see Lemma A.7.I.3).

<u>Proof</u>. $T = U^{-1}A$. If $\{\varphi_n\}$ is an arbitrary orthonormal basis of H then $\{U\varphi_n\}$ is an arbitrary orthonormal basis of H.

$$<TU\varphi_n, \varphi_n> = <U^{-1}AU\varphi_n, \varphi_n> = <AU\varphi_n, U\varphi_n> \quad .$$

Hence T is trace class. Q.E.D.

Corollary A.7.I.6. A necessary and sufficient condition for a bounded linear operator A on H to be trace class is that A^*A be compact and if $\{\varphi_n\}$ is an orthonormal basis of H so that $A^*A\varphi_n = \lambda_n\varphi_n$, then $\Sigma_{n=1}^{\infty} \lambda_n^{1/2} < \infty$.

Proof. $T\varphi_n = \lambda_n^{1/2}\varphi_n$. If A is trace class, then T is (Lemma A.7.I.5).

$$\text{tr } T = \sum \lambda_n^{1/2} .$$

If $\text{tr } T = \Sigma\lambda_n^{1/2} < \infty$, then T is summable (Lemma A.7.I.3). Hence $A = UT$ is trace class. Q.E.D.

If A is trace class and T is as above, define

$$||A||_1 = \text{tr } T .$$

Let $L_1(H)$ be the space of all trace class operators on H.

Lemma A.7.I.7. a) If $A,B \in L_1(H)$ then $A + B \in L_1(H)$ and $||A + B||_1 \leq ||A||_1 + ||B||_1$. $||\lambda A||_1 = |\lambda|\,||A||_1$.

b) $(L_1(H), ||\ ||_1)$ is a Banach space.

Proof. $A = US$, $B = VT$, S,T self adjoint and compact, U,V unitary and $\text{Tr}T = ||B||_1$, $\text{tr } S = ||A||_1$. Let $S\varphi_n = \lambda_n\varphi_n$, $T\psi_n = \mu_n\psi_n$, $\{\varphi_n\}$ and $\{\psi_n\}$ orthonormal bases. We note

$$||A||_1 = \sum \lambda_n = \sum ||A\varphi_n|| ,$$

$$||B||_1 = \sum \mu_n = \sum ||B\psi_n|| .$$

1) If $C \varepsilon L_1(H)$ and if $\{\xi_n\}$ is an orthonormal basis of H then

$$\|C\xi_n\| \leq \|C\|_1$$

Indeed, let $\{\eta_n\}$ be an orthonormal basis so that

$$\sum \|C\eta_n\| = \|C\|_1 .$$

$$\|C\eta_n\| = \Big\| \sum_\ell <\eta_n, \varphi_\ell> C\varphi_\ell \Big\|$$

$$\leq \sum_\ell |<\eta_n, \varphi_\ell>| \; \|C\varphi_\ell\|$$

Hence

$$\sum \|C\eta_n\| \leq \sum_{\ell,n} |<\eta_n, \varphi_\ell>| \; \|C\varphi_\ell\|$$

$$\leq \sum_\ell \left(\sum_n |<\eta_n, \varphi_\ell>|^2 \right)^{\frac{1}{2}} \|C\varphi_\ell\|$$

$$\leq \sum_\ell \|C\varphi_\ell\|$$

This proves 1).

To prove a) we note that

$$\sum_n \|(A+B)\xi_n\| \leq \sum_n \|A\xi_n\| + \|B\xi_n\|$$

$$\leq \|A\|_1 + \|B\|_1$$

b) Let $\{A_j\}$ be a Cauchy sequence in $L_1(H)$ relative to $\| \ \|_1$.

We note

2) If $A \in L_1(H)$, then $\|A\|_1 \geq \|A\|$. Indeed, if $\varphi \in H$ is a unit vector, then there is an orthonormal basis $\{\xi_n\}$ of H so that $\xi_1 = \varphi$. Then

$$\|A\varphi\| \leq \sum \|A\varphi_n\| \leq \|A\|_1 .$$

Hence $\|A\varphi\| \leq \|A\|_1$. Thus $\|A\| \leq \|A\|_1$.

2) implies that $\{A_j\}$ is Cauchy relative to $\|\ldots\|$. Hence there exists A_0 a bounded operator such that

$$\lim_{j\to\infty} A_j = A_0 ,$$

relative to $\|\ldots\|$.

Let $\epsilon > 0$ be given. Fix N: let j_0 be so large that $\|A_{j_0} - A_0\| < \epsilon/N$. Then if $\{\varphi_n\}$ is an orthonormal basis

$$\begin{aligned}\sum_{n\leq N} \|A_0\varphi_n\| &= \sum_{n\leq N} \|(A_0-A_{j_0})\varphi_n + A_{j_0}\varphi_n\| \\ &\leq \sum_{n\leq N} \|(A_0-A_{j_0})\varphi_n\| + \sum_{n\leq N} \|A_{j_0}\varphi_n\| \\ &< \epsilon + \|A_{j_0}\|\end{aligned}$$

Since $\{A_j\}$ is Cauchy relative to $||\ldots||_1$, $||A_j||_1 \leq C$ for all j. Hence

$$\sum_{n\leq N} ||A_0\varphi_n|| < \epsilon + C \quad .$$

Since ϵ and N are arbitrary, $A_0 \varepsilon L_1(H)$ and $||A_0||_1 \leq C$.

Let N be fixed as above. Let $\epsilon > 0$ be given. Let $\{\varphi_n\}$ be an orthonormal basis. Let j_0 be so large that if $j \geq j_0$, then $||A_j-A_0|| < \epsilon/N$. If $j \geq j_0$, then

$$\begin{aligned}\sum_n ||(A_j-A_0)\varphi_n|| &\geq \sum_{n\leq N} ||(A_j-A_0)\varphi_n|| \\ &\quad + \sum_{n>N} ||A_j\varphi_n|| + \sum_{n>N} ||A_0\varphi_n|| \\ &\leq \epsilon + \sum_{n>N} ||A_j\varphi_n|| + \sum_{n>N} ||A_0\varphi_n||\end{aligned}$$

(α) Given $\epsilon > 0$ and j_1 so that if $j,k \geq j_1$, $||A_j-A_k|| < \epsilon$ there exists N_1 so that if $j \geq j_1$, then

$$\sum_{n\geq N_1} ||A_j\varphi_n|| < \epsilon \quad .$$

In fact, let N_1 be such that $\Sigma_{n\geq N_1} ||A_{j_1}\varphi_h|| < \epsilon/2$. Now, if $j \geq j_1$,

$$\sum_{n\geq N_1} \|A_j\varphi_n\| \leq \sum_{n\geq N_1} \|(A_j - A_{j_1})\varphi_n\| + \sum_{n\geq N_1} \|A_{j_1}\varphi_n\|$$

$$< \epsilon/2 + \|A_j - A_{j_1}\|$$

$$= \epsilon .$$

Hence (α).

Fix $\epsilon > 0$ and j_1 as above. $j_0 \geq j_1$ and $N \geq N_1$. Then

$$\sum_n \|(A_j - A_0)\varphi_n\| < 2\epsilon + \sum_{n>N} \|A_0\varphi_n\| \quad .$$

Since N is arbitrary, we may take $\Sigma_{n>N} \|A_0\varphi_n\| < \epsilon$. This says that if $\{\varphi_n\}$ is an orthonormal basis, there is j_0 so that if $j \geq j_0$, then

$$\sum \|(A_j - A_0)\varphi_n\| < 3\epsilon \quad .$$

Since $\{\varphi_n\}$ is arbitrary we could start with j_1 so that if $k,\ell \geq j_1$, then $\|A_k - A_\ell\|_1 < \epsilon$ and if $\{\varphi_n\}$ so that

$$\|A_{j_1} - A_0\|_1 = \sum \|(A_{j_1} - A_0)\varphi_n\| \quad .$$

Hence

$$\|A_{j_1} - A_0\|_1 = \sum \|(A_{j_1} - A_0)\varphi_n\|$$

$$\leq \sum \|(A_{j_1} - A_0)\varphi_n\| + \sum \|(A_{j_0} - A_0)\varphi_n\|$$

$$< 4\epsilon \quad .$$

If $j \geq j_1$, then $\|A_j - A_0\|_1 \leq \|A_j - A_{j_1}\|_1 + \|A_{j_1} - A_0\|_1 < 5\epsilon$.

Q.E.D.

Corollary A.7.I.8 (to the proof of Lemma A.7.I.7). If $A \varepsilon L_1(H)$ then

$$\|A\|_1 = \sup_{\{\varphi_n\}} \sum \|A\varphi_n\|$$

the sup is taken over all orthonormal bases of H.

Lemma A.7.I.9. Let $k \varepsilon S(\mathbb{R}^2)$. Define for $\varphi \varepsilon S(\mathbb{R})$,

$$(K\varphi)(t) = \int_{-\infty}^{\infty} k(t,s)\varphi(s)\, ds \quad .$$

Then K extends to a bounded operator from $L^2(\mathbb{R})$ to $L^2(\mathbb{R})$. Furthermore, if $\{h_k\}$ is the orthonormal basis of $L^2(\mathbb{R})$ given by the Hermite functions (see A.5), then

$$\sum_{k,\ell} \langle Kh_k, h_\ell \rangle < \infty \quad .$$

Finally,

$$\operatorname{tr} K = \int_{-\infty}^{\infty} k(t,t)\, dt \quad .$$

<u>Proof</u>. Let

$$a_{k,\ell} = \int_{\mathbb{R}^2} k(t,s) h_k(t) h_\ell(s)\, dt\, ds$$

Then

$$a_{k,\ell} = \langle K h_k, h_\ell \rangle \quad .$$

Set

$$H_t = -\frac{\partial^2}{\partial t^2} + t^2 \quad ,$$

$$H_s = -\frac{\partial^2}{\partial s^2} + s^2 \quad .$$

Then as in the proof of Lemma 5.6.4, we see that if $p \geq 0$, $p \in \mathbb{Z}$,

$$(2k+1)^{-p}(2\ell+1)^{-p}|a_{k,\ell}| \leq \| H_t^p H_s^q k \| \quad .$$

This proves the first assertion of the lemma. As to the last,

$$\operatorname{tr} K = \sum_{n=1}^{\infty} a_{n,n} \quad .$$

Also using the results of A.5, we see

$$k(t,s) = \sum a_{n,m} h_n(t) h_m(s) \quad ,$$

the convergence as in $\mathcal{S}(\mathbb{R}^2)$. Hence

$$\int_{-\infty}^{\infty} k(t,t)\, dt = \sum a_{n,m} \int_{-\infty}^{\infty} h_n(t) h_m(t)\, dt$$

$$= \sum a_{n,n}$$

$$= \operatorname{tr} K \quad .$$

Q.E.D.

APPENDIX II TO CHAPTER 7

INDUCED REPRESENTATIONS

In this appendix we will study only the <u>most</u> elementary aspects of the theory of induced representations. For a more complete development, see Mackey [18], Bernat, et al. [2], or G. Warner [28], Chapter 5.

Let G be a unimodular Lie group and let $H \subset G$ be a closed unimodular subgroup. Fix on G (resp. H) a left (hence right) invariant measure dg (resp. dh). Fix on G/H a G-invariant measure $d(gH)$, so that

$$\int_G \varphi(g)\, dg = \int_{G/H} \left(\int_H \varphi(gh)\, dgh \right) d(gH)$$

for, say, $\varphi \in C_c^\infty(G)$.

If (ρ, V) is a unitary representation of H let $\mathcal{H}_0$ be the space of all $\varphi\colon G \to V$ so that

a) $\varphi(gh) = \rho(h)^{-1}\varphi(g)$, for $g \in G$, $h \in H$

b) $\|\varphi(g)\|^2 \in C_c(G/H)$. (Note $\|\varphi(gh)\| = \|\varphi(g)\|$ hence $\|\varphi(g)\|$ is a function on G/H.)

If $\varphi \in \mathcal{H}_0$, define

$$\|\varphi\|^2 = \int_{G/H} \|\varphi(g)\|^2\, dg \quad .$$

Let $< , >$ denote the corresponding pre-Hilbert space structure on $\mathcal{H}_0$.

If $\varphi \in \mathcal{H}_0$ and $g \in G$, define

$$(T(g)\varphi)(X) = \varphi(g^{-1}X) \quad .$$

Then

$$T(g): \mathcal{H}_0 \to \mathcal{H}_0 \quad .$$

Also, it is easy to see that

$$||T(g)\varphi|| = ||\varphi||$$

$$\cdot \varphi \in \mathcal{H}_0 \quad .$$

Let $\mathcal{H}$ denote the Hilbert space completion of $\mathcal{H}_0$. Then, by the above, $T(g)$ extends to a unitary operator from $\mathcal{H}$ to $\mathcal{H}$.

<u>Lemma A.7.II.1</u>. $(T,\mathcal{H})$ is a unitary representation of G denoted $\mathrm{Ind}_H^G(\rho)$.

For a proof, see, for example, Wallach [26], Chapter 2.

<u>Lemma A.7.II.2</u>. Let G be a unimodular Lie group and let $H_1 \subset G$ be a closed subgroup so that H_1 is unimodular. Let $H_2 \subset H_1$ be a closed subgroup of H_1 so that H_2 is unimodular. If (ρ,V) is a unitary representation of H_2, then $\mathrm{Ind}_{H_2}^G(\rho)$ is unitarily equivalent with $\mathrm{Ind}_{H_1}^G(\mathrm{Ind}_{H_2}^{H_1}(\rho))$.

Proof. Set

$$\mathrm{Ind}_{H_2}^{H_1}(\rho) = (\pi_1, \mathcal{H}_1) \quad ,$$

$$\mathrm{Ind}_{H_2}^{G}(\rho) = (\pi_2, \mathcal{H}_2)$$

and

$$\mathrm{Ind}_{H_1}^{G}(\pi_1) = (\delta, \mathcal{H}) \quad .$$

Let $\mathcal{H}_{1,0}, \mathcal{H}_{2,0}$ and $\mathcal{H}_0$ be the corresponding pre-Hilbert spaces of the definitions. If $f \in \mathcal{H}_{2,0}$, define

$$(Af)(g)(h_1) = f(gh_1)$$

$$\text{for } g \in G, \; h_1 \in H_1 \quad .$$

If $g \in G$ is fixed and $h_1 \in H_1$, $h_2 \in H_2$, then

$$\begin{aligned} Af(g)(h_1h_2) &= f(gh_1h_2) \\ &= \rho(h_2)^{-1}(Af)(g)(h_1) \quad . \end{aligned}$$

If $h_1, x \in H_1$, then

$$\begin{aligned} (Af)(gx)(h_1) &= f(gxh_1) \\ &= (Af)(g)(xh_1) \\ &= (\pi_1(x)^{-1}((Af)(g)))(h_1) \quad . \end{aligned}$$

Hence $Af(g) \in \mathcal{H}_{1,0}$ and $Af\colon G \to \mathcal{H}_{1,0}$ is in $\mathcal{H}_0$. Now,

$$\begin{aligned}
\|Af\|^2 &= \int_{G/H_1} \|(Af)(g)\|^2 \, d(gH_1) \\
&= \int_{G/H_1} \int_{H_1/H_2} \|Af(g)(h_1)\|^2 \, d(h_1H_2)\, d(gH_1) \\
&= \int_{G/H_1} \int_{H_1/H_2} \|f(gh_1)\|^2 \, d(h_1H_2)\, d(gH_1) \\
&= \int_{G/H_2} \|f(g)\|^2 \, d(gH_2) \quad .
\end{aligned}$$

Hence $A\colon \mathcal{H}_{2,0} \to \mathcal{H}_0$ is unitary. Let A also denote its unitary extension from $\mathcal{H}_2 \to \mathcal{H}_0$. Clearly, $A \circ \pi_2(g) = \delta(g) \circ A$.

If $f \in \mathcal{H}_0$, define

$$(Bf)(g) = f(g)(1) \quad .$$

Then $Bf \in \mathcal{H}_{2,0}$ and $\|Bf\| = \|f\|$. Clearly, $BA = I$ and $AB = I$ (B denotes the unitary extension of B to $\mathcal{H}$). Q.E.D.

Bibliography

1. L. Auslander and B. Kostant, "Polarization and Unitary Representations of Solvable Lie Groups", Invent. Math. 14 (1971), 155-254.

2. P. Bernat, et al., Representations des Groupes de Lie Résolubles, Dunod, Paris, 1972.

3. R. Blattner, "Quantization and Representation Theory", Proc. of Symp. in Pure Math., XXVI, Amer. Math. Soc., Providence, Rhode Island, 1973, 147-165.

4. J. Brezin, Unitary Representation Theory for Solvable Lie Groups, Mem. Amer. Math. Soc., No. 79 (1968).

5. P. Cartier, "Quantum Mechanical Commutation Relations and Theta Functions", Proc. of Symp. in Pure Math., IX, Amer. Math. Soc., Providence, Rhode Island, 1966, 361-386.

6. J. Dixmier, "Sur les Représentations Unitaires des Groupes de Lie Nilpotents", I, II, III, IV, V, VI, Amer. J. Math. 81 (1959), 325-388; Canad. J. Math. 10 (1958), 321-348; Canad. J. Math. 11 (1959), 321-344; Bull. Soc. Math. France 87 (1959), 65-79; Canad. J. Math. 12 (1960), 324-352.

7. J. Dixmier, Algébres Enveloppants, Gauthier-Villars, Paris, 1974.

8. S. Helgason, Differential Geometry and Symmetric Spaces, Academic Press, New York, 1962.

9. R. Hermann, Differential Geometry and the Calculus of Variations, Academic Press, New York, 1968.

10. F. Hirzbruch, Topological Methods in Algebraic Geometry, Springer-Verlag, New York, 1960.

11. J. Igusa, Theta Functions, Springer-Verlag, Berlin, 1972.

12. N. Jacobson, Lie Algebras, Interscience, New York, 1962.

13. A. Joseph, "Derivations of Lie Bracket and Canonical Quantization", Commun. Math. Phys. 17 (1970), 210-232.

14. A.A. Kirillov, "Unitary Representations of Nilpotent Lie Groups", Russ. Math. Surveys 17 (1962), 53-104.

15. S. Kobayashi and K. Nomizu, Foundations of Differential Geometry, Vols. I, II, Interscience, New York, 1962, 1969.

16. B. Kostant, "Quantization and Unitary Representations", Lecture Notes in Math., Vol. 170, Springer-Verlag, Berlin, 87-208.

17. S. Lang, Analysis II, Addison-Wesley, Reading, 1969.

18. G.W. Mackey, "Induced Representations of Locally Compact Groups, I", Ann. of Math. 55 (1952), 101-139.

19. J. Milnor, "Morse Theory", Annals of Math. Studies, No. 51, Princeton University Press, Princeton, 1963.

20. R. Palais, "A Global Formulation of the Lie Theory of Transformation Groups", Mem. of the Amer. Math. Soc., No. 22, Amer. Math. Soc., Providence, Rhode Island, 1957.

21. C. Pukansky, Lecons sur les Représentations des Groupes, Monographics de la Soc. Math. France, No. 2, Dunod, Paris, 1967.

22. D. Shale, "Linear Symmetries of Free Boson Fields", Trans. Amer. Math. Soc. 103 (1962), 149-167.

23. B. Simon, article in Mathematics of Contemporary Physics, Edited by R.F. Streator, Academic Press, London, England, 1972.

24. J. Souriau, Structure des Systemes Dynamiques, Dunod, Paris, 1970.

25. L. van Hove, "Sur Certaines Representations d'un Groupe Infine de Transformations", Acad. Roy. Belg. Cl. Sci. Mem. Coll. 26 (1951), 102-198.

26. N.R. Wallach, Harmonic Analysis on Homogeneous Spaces, Marcel Dekker, New York, 1973.

27. F. Warner, Foundations of Differential Geometry and Lie Groups, Scott, Foresman and Co., Glenview, 1971.

28. G. Warner, Harmonic Analysis on Semisimple Lie Groups, I, II, Springer-Verlag, Berlin, 1972.

29. A. Wiel, "Sur les Théorèmes de de Rham", Comm. Math. Helv. 26 (1952), 119-145.

30. A. Wiel, Variétés Kählériennes, Hermann, Paris, 1958.

31. A. Wiel, "Sur Certaines Groupes d'Opérateurs Unitaries", Acta Math. 111 (1964), 143-211.

32. A. Weinstein, "Symplectic Manifolds and Their Lagrangian Submanifolds", Advances in Math. 6 (1971), 329-346.

33. N. Wiener, The Fourier Integral and Certain of its Applications, Dover, New York (1963 edition).

APPENDIX ON QUANTUM MECHANICS

by Robert Hermann

1. INTRODUCTION

One of the prime motivating factors in the study of symplectic manifolds is the role they play in classical mechanics. (This relation is very old. It certainly is implicitly in the work of Hamilton and Jacobi, and it became explicit in Lie's work on contact transformations and "function groups" and finally takes almost the modern form in Poincaré's and Cartan's work on "invariant integrals".) It has recently been realized that they also play an important role in Lie group representation theory--of course, this is the main theme of this book--*and* in the theory of asymptotic behavior of solutions of linear partial differential equations (the work of J. Keller, Maslov, Hörmander...)

Is this appearence of the same mathematics in different areas an accident? Surely not. In fact, the existence of these relations should reflect the existence of *quantum mechanics*. However, the relations between quantum mechanics and symplectic manifold theory are much less clear-cut and direct. In fact, there are at least two interconnections. First, quantum mechanics is related to classical mechanics via the various "quantization" procedures (Bohr-Sommerfeld, De Broglie, Heisenberg,

Schrödinger, Weyl, Dirac,...) and the "correspondence principle" (that quantum mechanics should, in _some_ sense, reduce to classical as Planck's constant goes to zero). Thus, one might think that there is a _direct_ connection between quantum mechanics and symplectic manifolds, which one can study _independently of classical mechanics_. Second, quantum mechanics shares with classical mechanics the property that its state space is a symplectic manifold. (In the classical mechanics of particles with "n degrees of freedom", the state space is, at least locally, R^{2n}, with its usual symplectic structure. For quantum mechanics, it is the projective space associated to a Hilbert space. Since the Hilbert space will usually be infinite dimensional, the state space will be an "infinite dimensional symplectic manifold"--objects for which there really is as yet no general theory.)

Another reason for looking for links between symplectic manifold theory and quantum mechanics might be to better understand the reason that symplectic manifolds seem to be important for Lie group theory. It seems to be a characteristic of quantum mechanics that group theory plays a much more decisive role than in classical mechanics. Again, there are many facets to this observation. (For example, the only physical systems which one can exhibit as quantum systems with a reasonable degree of completeness and precision are those--the hydrogen atom and harmonic oscillator, and certain linear wave equations like the Klein-Gordon and Dirac equation--which admit a high degree of "symmetry". To a certain extent, this emphasis on group theory in quantum mechanics goes beyond strictly mathematical and logical reasons. For example, in elementary particle physics, experiments seem to suggest that "elementary particles"--and we do not really know what they are--obey certain rules in their interactive and stability behavior which have a group theoretic foundation. (As an analog, think of the fact that one can perceive certain common themes of "symmetry", i.e., "group theory", in art and architecture, throughout all periods of human history, and relatively independently of the specific techniques and mediums used in that art.) Now, one definite reason for this relation to play a prominent role is that in _both_ quantum and classical mechanics there is a correspondence between one-parameter group of "symmetries", i.e., automorphisms of the structure, and _conservation laws_. (In fact, this correspondence is a manifestation of the fact that the state space is a symplectic manifold.)

At any rate, it should be clear that there are complicated and immensely rich (but badly understood) relations between classical and quantum mechanics (hence also all those branches of physics which are strongly dependent for their foundations on classical or quantum mechanics, such as continuum mechanics, atomic, nuclear and sub-nuclear physics, statistical mechanics, etc.), Lie group theory and symplectic manifold theory. One obstacle to wider work on these problems is the lack of a reasonably definitive, modern treatise on the mathematical structure of quantum mechanics. (What does exist is mainly--following the classic by von Neumann--devoted to the functional analysis side, which is not so directly relevant to these issues.) In particular, there is really no place that a mathematician can learn quantum mechanics in a reasonably efficient way. Since I have been rash enough to publish several books and to lecture at various universities about these topics, Professor Wallach has suggested that I write this appendix to supplement his treatment of the differential geometry and Lie group theory of symplectic manifolds and to help guide the (mathematical) reader into these wider pastures.

I am afraid that I cannot hope to be understood by a mathematician reader who starts from zero knowledge of physics. For such a person, I suggest some preliminary reading. Feynman's "Lectures on Physics" [1] are probably the best general introduction to current physical thought for scientists in other disciplines. Goldstein [1] and Corben-Stehle [1] are by now traditional textbooks on classical mechanics for physics students which are moderately accessible to mathematicians. Abraham and Marsden's "Foundations of Mechanics" [1] is probably the most successful recent book in teaching mechanics to mathematicians. (Of course, it is most valuable from our point of view because of its emphasis on the relation between mechanics and symplectic manifolds.)

Equivalent material for quantum mechanics does not exist, and learning quantum mechanics is, for most mathematicians, a painful business. Of course, most mature mathematicians will be familiar with the elements of functional analysis, and will like parts of treatises which approach quantum mechanics from that point of view, e.g., von Neumann [1], Prugovecki [1] and Reed-Simon [1]. Mackey's Lectures [1] are probably still the best, and the closest one can come to the example of Abraham-Marsden. However, they will probably not learn very well from these books (except possibly Mackey, who does try to give some insights into the physical motivation) how physicists think about quantum mechanics! For this, the only sources available are the standard textbooks for a graduate level physics course in quantum mechanics. (The more elementary books are usually more painful for a mathematician--either they lapse into baby talk about the mathematics, or else they have a very formalistic tone which is a pale shadow and a poor substitute for von Neumann). The ones I found most useful are by Gottfried [1] and Messiah [1], although there is nothing particularly sacred about this choice.

2. A MATHEMATICAL MODEL FOR CLASSICAL MECHANICS

For a mathematician, "classical mechanics" means, roughly, Newton's Laws

$$\text{Force} = \text{Mass} \times \text{Acceleration}$$

The physics enters at the level of understanding what "force", "mass", and "acceleration" mean, how they are used to describe real-world situations, how they combine in interaction, etc. All of this can be described quite completely within the realm of ordinary differential equations. (Whittaker's treatise [1] is still, for my money, the best book on classical mechanics.) Of course, for two hundred years after Newton this was the prime motivation for studying differential equations!

Nowadays, many applied mathematicians would say that this is a matter of mathematical modelling, and that classical mechanics seems to be the theory of systems that can be modelled by means of certain types of ordinary differential equations. Now, there is something a bit too glib, and perhaps too technocratic and even unscientific (taking science at its highest level) about this approach--for example, it tends to equate the scientific level of Newton's laws with the model your friendly neighborhood economist uses to (mis) predict the U.S. economy in the next year. However, it will serve our immediate purposes.

Thus, a suitable model for a classical mechanical system is a pair (M,X) consisting of a manifold M and a vector field X on M. (This sort of model is also hypothesized by R. Thom in his poetic unification of diverse phenomenon in terms of "catastrophes" [1].) M is typically defined by the "kinematics", i.e., the positions, velocities, masses and constraints on the particles which make up the system. X is determined by the "dynamic" data, e.g., the forces. The integral curves of X, i.e., the curves $t \to p(t)$ in M such that

$$\frac{dp}{dt} = X(p(t)) \quad , \tag{2.1}$$

determines the time evolution or dynamics of the system.

Let us suppose, for simplicity, that the forces, constraints, masses, etc., are free of explicit time dependence. Then, the differential equations (2.1) determining the time evolution of the system are autonomous. Group theoretically, this means that the Lie group R (i.e., the group isomorphic to the additive group of the real numbers) acts as a transformation group on M. Thus, at a very abstract level, one might say that a suitable mathematical model for the study of mechanics is a pair consisting of a manifold M, and a transformation group action of R on M. One would expect that quantum mechanics (and much more besides!) can be described also by such a model.

Now, my plan for this Appendix is to abstract from classical mechanics three key ideas--conservation laws, state and observables. I will show how they can be given a quasi-algebraic structure which is intimately related to the fact that the manifold M carries a symplectic geometric structure. Then, I will show how the most commonly used "model" for quantum mechanics--with Hilbert spaces replacing symplectic manifolds--reproduces this algebraic structure. Thus, we now ascend to a higher level of abstraction (and "generalized nonsense"!), returning later to the concrete situations of interest in physics.

3. CONSERVATION LAWS, STATE AND OBSERVABLES FOR DYNAMICAL SYSTEMS

As explained above, I want to develop a setting general enough to encompass both for classical and quantum mechanics certain ideas that are usually only implicit in the physicist's treatment. (Roughly, they involve what the physicist calls the Heisenberg and Schrödinger "pictures".

Let M be a space. Let t denote a real parameter, varying over

$$-\infty < t < \infty \quad .$$

Let t denote the space of this parameter. Suppose given a one parameter group

$$t \to \phi_t$$

of transformations on M. (This means that:

a) Each ϕ_t is a map: $M \to M$

b) $\phi_{t_1+t_2} = \phi_{t_1}\phi_{t_2}$ for $t_1,t_2 \in R$

c) ϕ_0 = identity map

The pair (M,ϕ) is called a <u>dynamical system</u>.

Definition. A _state_ of the dynamical system is a curve

$$\sigma: T \to M$$

which is an orbit of ϕ, i.e., such that

$$\sigma(t) = \phi_t(p)$$

for some $p \in M$

The _state space_ of the dynamical system is the set of all such states.

Remark. In this situation, the "states" of the system are, of course, identified with M itself. In more complicated situations--and in field theories--this identification will break down. I will make some comments about more general situations later on.

Definition. A _conservation law_ for the dynamical system is a map

$$\gamma: M \times T \to R$$

such that:

$$t \to \gamma(\phi_t(p),t) = \text{constant} \tag{3.1}$$

for all $p \in M$.

(In words, whenever $t \to p(t)$ is an orbit curve of the group, $t \to \gamma(p(t),t)$ is constant along that curve.

<u>Definition</u>. An <u>observable</u> for the dynamical system is a real-valued function

$$\underline{f}\colon \text{(state space)} \to R$$

Define a map

$$\alpha\colon \text{(conservation laws)} \to \text{(observables)} \tag{3.2}$$

as follows:

$\alpha(\gamma)$ assigns to the curve $t \to \sigma(t)$ which is an orbit of ϕ the constant value of $t \to \gamma(p(t),t)$

(3.3)

Now, we can let the group T act on the set of conservation laws as follows:

$$\phi_{t_0}(\gamma)(p,t) = \gamma(p, t - t_0) \tag{3.4}$$

Let us verify that for each $t_0 \in T$ this formula defines $\phi_{t_0}(\gamma)$ as a conservation law. Let

$$t \to p(t) = \phi_t(p_0)$$

be an orbit curve. Then,

$$\begin{aligned} \phi_{t_0}(\gamma)(p(t),t) &= \gamma(p(t),\ t-t_0) \\ &= \gamma(\phi_t(p_0),\ t-t_0) \\ &= \gamma(\phi_{t-t_0}(\phi_{t_0}p_0),\ t-t_0) \quad , \end{aligned}$$

using the group property for ϕ_t. As a function of t, this is constant, since $s \to \phi_s(\phi_{t_0}p_0)$ is also an orbit curve, i.e., a "state".

We can now put together formulas (3.3) and (3.4):

$$\begin{aligned} &\alpha(\phi_{t_0}(\gamma)) \text{ has a value on the state} \\ &t \to p(t) \text{ the real number } \gamma(p(t_0),0) \end{aligned} \qquad (3.5)$$

Let us consider the physical meaning of these objects in a simple situation. Suppose that $M = R^2$ represents the position and velocity coordinates of a particle moving along the real line. Then, a "state" is a curve

$$t \to p(t) = (x(t),\ v(t))$$

which satisfies Newton's equation:

$$\begin{aligned} \frac{dx}{dt} &= v(t) \\ m\frac{dv}{dt} &= F(x(t),\ v(t)) \end{aligned} \qquad (3.6)$$

(In order to obtain the group property for ϕ_t, notice that the "force" function $F(\, , \,)$ must be a <u>function of</u> x,v alone, i.e., "time-independent".) x is the "position", v the "velocity

Let

$$\gamma(p,t) = \text{position coordinate at time } t = 0 \text{ of the particle which is at } p \text{ for time } t. \tag{3.7}$$

(Notice that, by its definition, γ is "conserved".) Thus:

$\alpha(\gamma)$ assigns to the curve $t \to p(t)$ the position corrdinate of the particle at time $t = 0$. (3.8)

$$\begin{aligned} \alpha(\phi_{t_0}(\gamma))(\{p(t)\}) &= \gamma(p(t_0),0) \\ &= \text{position coordinate of particle at time } t = t_0 \end{aligned} \tag{3.9}$$

Now, return to the general setting. Given a conservation law

$$\gamma \colon M \times T \to R \quad ,$$

set:

$$f(p) = \gamma(p,0) \quad . \tag{3.10}$$

f is then a real-valued function $M \to R$. Thus, from (3.5) we have:

Value of the observable $\alpha(\phi_t(\gamma))$ on the state $\{t \to p(t)\}$ is $f(p(t))$.

(3.11)

Conversely, a map

$$f: M \to R$$

arises in this way from a conservation law γ via the formula:

$$\gamma(p,t) = f(\phi_{-t}(p)) \tag{3.12}$$

Thus, we have a set of maps summed up in the following diagram:

(real-valued functions on M)
↗
(conservation laws) (3.13)
↘
(curves in the space of observables)

Remarks. The difference between the observables $\alpha(\gamma)$ and $\alpha(\phi_t(\gamma))$ is an abstraction of what is called the *Lagrangian* and *Eulerian picture* in continuum mechanics, and the *Schrödinger* and *Heisenberg picture* in quantum mechanics.

Notice that these definitions are quite general. (Of course, one reason for this is that they are quite trivial!) For example, the group T can be replaced by an arbitrary group G which acts as a transformation group on M.

They also have great generality and flexibility in a differential-geometric direction. For example, here is how they may be formulated within the framework of E. Cartan's theory of Exterior Differential Systems:

I is a differential ideal of forms on a manifold E.

N is a fixed manifold of dimension n, with boundary ∂N.

The conservation laws are the $(n-1)$-forms θ on E such that

$$d\theta \in I \quad .$$

The states are the submanifold maps

$$\beta: N \to E$$

such that

$$\beta^*(I) = 0 \quad .$$

Observables are real-valued functions on states.

The map

$$(\text{conservation laws}) \to (\text{observables})$$

is defined by:

$$\theta \text{ assigns to } \beta \text{ the number } \int_{\partial N} \beta^*(\theta) \quad .$$

4. OBSERVABLES AND CONSERVATION LAWS FOR ONE-PARAMETER DIFFEOMORPHISM GROUPS ON SYMPLECTIC MANIFOLDS

So far, the exact category of mappings within which the definitions of Section 3 are to make sense has not been specified.

"Classical" and "quantum" mechanics lead to quite distinctive choices. Let us examine the situation in more detail for the case where M is a manifold, and $t \to \phi_t$ is a one parameter group of diffeomorphisms. Let X be the vector field on M which is the infinitesimal generator of this group. Then,

$$X(f)(p) = \frac{d}{dt} f(\phi_{-t}p)\Big|_{t=0} \tag{4.1}$$

for each C^∞ real valued function on M. We can also write (4.1) as:

$$\frac{dp}{dt} = X(p(t)) \quad , \tag{4.2}$$

where $t \to \phi_t(p)$ is an orbit curve of the group. (dp/dt denotes its tangent vector.)

A map

$$\gamma: M \times T \to R$$

is then a conservation law if

$$0 = \frac{d}{dt}\gamma(p(t),t) = X(\gamma)(p,t) + \frac{\partial\gamma}{\partial t}(p(t),t) \quad .$$

Hence, the condition that γ be conserved is:

$$\frac{\partial\gamma}{\partial t} + X(\gamma) = 0 \quad . \tag{4.3}$$

Thus, given $f\colon M \to R$, finding a γ such that f is assigned to γ via formula (3.9) is equivalent to solving the "Cauchy problem" for the first order partial differential equation (4.3).

Now, suppose that M is a symplectic manifold. In the text, it has been shown that $C^\infty(M)$ (the space of C^∞ real valued functions on M) has a Lie algebra structure, called Poisson bracket, denoted by

$$\{f_1, f_2\} \quad .$$

If γ_1, γ_2 are maps: $M \times T \to R$, let

$$\{\gamma_1, \gamma_2\}$$

denote the function of (p,t) obtained by holding t constant, and calculating the Poisson bracket of the elements

$$p \to \gamma_1(p,t)$$

$$p \to \gamma_2(p,t)$$

of $C^\infty(M)$.

Let ω be the closed two-form on M which defines the symplectic structure. Suppose that the one-parameter group $t \to \phi_t$ consists of symplectic automorphisms. The condition for this is readily seen to be that the Lie derivative of ω by the vector field X is zero, i.e.,

$$\mathscr{L}_X(\omega) = 0 \quad . \tag{4.4}$$

<u>Theorem 4.1</u>. If (γ_1,γ_2) are two conservation laws for the symplectic automorphism dynamical system $\{\phi_t\}$, then

$$\{\gamma_1,\gamma_2\}$$

is also a conservation law. In other words, the set of conservation laws admits a Lie algebra structure, via the Poisson bracket.

<u>Proof</u>. We must show that

$$\frac{\partial}{\partial t}\{\gamma_1,\gamma_2\} + X(\{\gamma_1,\gamma_2\}) = 0$$

This follows by a routine calculation that is left to the reader.

Now, consider the one-parameter group

$$\gamma \to \phi_t(\gamma) \tag{4.5}$$

acting on the conservation laws, defined by formula (3.4).

<u>Theorem 4.2</u>. The group transformations (4.5) are automorphisms of the Poisson bracket Lie algebra structure on the space of conservation laws.

<u>Proof</u>. Equation (4.3) shows that Lie derivation by X is the infinitesimal generator of the group (4.5). It acts as a derivation of the Poisson bracket structure, because of (4.4).

Theorem 4.3. The map

$$\gamma \to f$$

defined by formula (3.9) of

$$(\text{conservation laws}) \to C^{\infty}(M)$$

is an isomorphism of the Poisson bracket Lie algebra structures.

Proof. The proof is left to the reader.

Theorems 4.1 - 4.3--almost trivial in the context in which we are working--are the key ingredients we would like to carry over to a theory of "observable-conservation law-states" for more general dynamical systems. We shall see that the theory of Hilbert spaces provides the simplest route to do this for quantum mechanics.

We shall now use the map

$$(\text{conservation laws}) \to (\text{observables})$$

(which in this case is a vector-space isomorphism) to make the set of observables into a Lie algebra, so that this map becomes a Lie algebra isomorphism. In words, if $(p,t) \to \gamma_1(p,t), \gamma_2(p,t)$ are conservation laws, and if $\underline{f}_1, \underline{f}_2$ are the corresponding observables, i.e.,

$$\underline{f}_1 \text{ (on the curve } t \to p(t)) = \gamma_1(p(0),0)$$

$$\underline{f}_2 \text{ (on the curve } t \to p(t)) = \gamma_2(p(0),0) \quad ,$$

then the Poisson bracket $\{\underline{f}_1,\underline{f}_2\}$ is defined so that

$$\{\gamma_1,\gamma_2\} \to \{\underline{f}_1,\underline{f}_2\} \quad .$$

Having defined the "conservation laws", we have the map

$$\gamma \to \text{(curve in the space of observables)}$$

defined in the abstract context by formula (3.5). Let

$$t \to \alpha(\phi_t(\gamma)) \quad \equiv \quad \underline{f}_t$$

denote this curve. It satisfies a differential equation related to (4.3). We shall examine it for the simplest situation of mechanics, a particle of mass m moving on a line (with coordinate x) under the influence of potential function $V(x)$.

M is then

$$R^2 \quad = \quad \text{space of variables} \quad (x,y)$$

(x stands for <u>position</u>, y for <u>momentum</u>.) Set:

$$H(x,y) \quad = \quad \frac{1}{2m}\, y^2 + V(x) \quad . \tag{4.6}$$

Let the (symplectic automorphism) dynamical system be defined by the Hamilton equations:

$$\begin{aligned} \frac{dx}{dt} &= \partial_y H = \frac{y}{m} \\ \frac{dy}{dt} &= -\partial_x H = -V'(x) \end{aligned} \tag{4.7}$$

Together, equations (4.7) are, of course, equivalent to Newton's second order equations:

$$m \frac{d^2x}{dt^2} = -V'(x) \quad . \tag{4.8}$$

Let γ^x, γ^y be the conservation laws defined as follows:

$$\begin{aligned} \gamma^x(x,y,t) &= x \text{ - component of } \phi_{-t}(x,y) \\ \gamma^y(x,y,t) &= y \text{ - component of } \phi_{-t}(x,y) \end{aligned} \tag{4.9}$$

Exercise. Show that equations (4.3) take the following form:

$$\begin{aligned} \frac{\partial}{\partial t} \gamma^x &= \{H, \gamma^x\} \\ \frac{\partial}{\partial t} \gamma^y &= \{H, \gamma^y\} \end{aligned} \tag{4.10}$$

Now, $H(x,y)$ is itself a conservation law, since:

$$\frac{d}{dt} H(x(t), y(t)) = 0$$

along solutions of (4.7). (Physically, this is just "conservation of total energy.") Let

$$\underline{x}^t = \text{observable corresponding to } \phi_t(\gamma^x)$$

$$\underline{y}^t = \text{observable corresponding to } \phi_t(\gamma^x)$$

Set:

$$\underline{H} = H(\underline{x}^t, \underline{y}^t) \quad .$$

It too is an observable. Corresponding to (4.10), we then have:

$$\frac{d}{dt}\underline{x}_t = \{\underline{H}, \underline{x}_t\}$$

$$\frac{d}{dt}\underline{y}_t = \{\underline{H}, \underline{y}_y\} \qquad (4.11)$$

These are the <u>Heisenberg equations of motion of observables</u>. Exactly in the way we derived Newton's equations (4.8) for the <u>curves</u>, the Heisenberg equations take the form:

$$m \frac{d^2 \underline{x}}{dt^2} = -V'(\underline{x}) \qquad (4.12)$$

Of course, the mathematical meaning of the "unknowns" $t \to x(t)$, $t \to \underline{x}(t)$ in (4.8) and (4.12) are completely different. In (4.12), each $\underline{x}(t)$ is to be a real-valued function on a space. $V'(\underline{x})$ is to be constructed using the <u>commutative associative algebra structure</u> on the space of such functions, provided by addition and multiplication in the ordinary sense.

<u>Remark</u>. I cannot resist pointing out how close (4.12) is to quantum mechanics. To obtain it, consider equations (4.12), where, however,

$$t \to \underline{x}(t)$$

is to be a curve in the space of self-adjoint operators in a Hilbert space H, with "initial condition" commutative relations:

$$\frac{2\pi m}{ih}\left[\frac{d\underline{x}}{dt}(0), \underline{x}(0)\right] = 1 \quad , \tag{4.13}$$

where h = Planck's constant. The mathematician reader can now appreciate why physicists are notoriously unwilling to worry about the mathematical meaning of quantum mechanics! They are "pragmatic", and can get the results they need by replacing in Newton's equations the real-valued functions of t by self-adjoint operator valued functions of t, subject to "canonical commutation relations" (4.13). This enables them to reverse the natural order of classical mechanics; first define the "observables", "represent" them in a suitable way (in this case, as self-adjoint operators on a Hilbert space), then define the "states" as the objects on which the observables act. In classical mechanics, the "states" are always considered first, as points on a manifold.

However, their mathematical "pragmatism" catches up with them when it comes to quantum field theory. There, they attempt to replace the equations of motion for "classical" field (regarding "classical" field as, say, a map $R^4 \to$ (a real vector space)) by operator-valued functions on R^4. Setting up equations analogous to (4.12) requires them to multiply such objects. The analogue of the commutative relation (4.12) requires that they not be operator-valued _functions_ but operator-valued _distributions_. (The tip off is that Dirac delta functions appear on the right hand side of the generalization of (4.13).) These, it turns out, are completely contradictory requirements! There is no reasonable way to _consistently_ multiply operator-valued distributions. This does not stop them--they devise ingenious ways of making some sense of things--usually at the expense of introducing perturbation series which never converge, and, in fact, whose _terms_ usually have the value infinity anyway! Still, they do not stop--they plunge on and devise a procedure called "renormalization" to throw away these infinities.

It is also worth pointing out another direction in which the ideas can be generalized. Consider an arbitrary manifold and an _arbitrary_ system of ordinary differential equations on M--not necessarily arising from a vector field. Thus, symbolically, it will be of the form

$$D\left(p, \frac{dp}{dt}, \frac{d^2p}{dt^2}, \ldots\right) = 0 \quad .$$

A _state_ can now be defined as a curve $t \to p(t)$ in M which is a solution of the differential equation. A _conservation law_ is a mapping

$$\gamma : \left(p, \frac{dp}{dt}, \frac{d^2p}{dt^2}, \ldots, t\right) \to R$$

which is constant on the solutions of the differential equation. An _observable_ is again a function

$$(\text{state}) \to R \quad .$$

We should now like to put some sort of Lie algebra structure on the observables. How to do this in a reasonably general case is not at all obvious--in fact, has never been treated in adequate generality. _If_ the differential equations are the Euler-Lagrange equations of a calculus of variations problem--and if the Lagrangian satisfies certain "non-singularity" conditions--this can be done. My books listed in the bibliography contain much information about this area.

Before turning to quantum mechanics in its full glory, let us consider a specialization of the situation in this chapter--to systems determined by linear differential equations.

5. CONSERVATION LAWS, STATES AND OBSERVABLES FOR LINEAR SYMPLECTIC DYNAMICAL SYSTEMS

Whenever the space M on which the "dynamical" group $t \to \phi_t$ acts has a special structure--which is preserved by each ϕ_t--it is convenient to think of choosing the conservation laws to be compatible with that structure. We shall now consider such a situation. In physics, it occurs for "classical" systems whose equations of motion are _linear_. It is one of the most curious facts in all of science that an infinite dimensional version of it describes <u>quantum</u> systems--even those whose classical equations of motion are nonlinear.

Since we shall be dealing in this section mainly with vector spaces, we shall change the notation slightly to conform more closely with standard linear algebra.

Let V be a real vector space. (Assume, for the moment, that it is finite dimensional.) Let

$$\omega: V \times V \to R$$

be a skew-symmetric, bilinear, non-degenerate form. The pair

$$(V,\omega)$$

is said to be a <u>linear symplectic space</u>. A linear map

$$\phi: V \to V$$

is called a <u>symplectic automorphism</u> if

$$\omega(\phi(v_1), \phi(v_2)) = \omega(v_1,v_2)$$

for $v_1,v_2 \in V$.

A linear map

$$A: V \to V$$

is an <u>infinitesimal symplectic automorphism</u> if

$$\omega(Av_1,v_2) + \omega(v_1,Av_2) = 0 \tag{5.1}$$

for $v_1,v_2 \in V$.

Suppose that

$$t \to \phi_t$$

is a one-parameter group of symplectic automorphisms of V, with A as infinitesimal generator. The orbits of ϕ then satisfy the following equation:

$$\frac{dv}{dt} = Av \quad . \tag{5.2}$$

We can then apply the general formalism of Section 3--define states, observables, conservation laws. Since we are endowing the dynamical group with a special structure, it is reasonable to consider special classes of conservation laws. In fact, we shall consider those that are <u>quadratic</u> in v, i.e., of the form

$$\gamma(v,t) = \omega(B(t)v, v) \quad , \tag{5.3}$$

where each $B(t)$ is a linear map $V \to V$. We can suppose without loss in generality that

$$\begin{aligned} \omega(Bv_1,v_2) &= \omega(Bv_2,v_1) \\ &= -\omega(v_1,Bv_2) \quad , \end{aligned} \tag{5.4}$$

i.e., that each $B(t)$ is itself an infinitesimal symplectic automorphism, i.e., satisfies (5.1).

The condition that (5.3) define a conservation law for equation (5.2) is that:

$$\frac{d}{dt}\,\gamma(v(t),\ t) \;=\; 0 \tag{5.5}$$

for all solutions of (5.2),

i.e., that

$$\omega\left(\frac{dB}{dt}\,v,\ v\right) + \omega(BAv,v) + \omega(BV,Av) \;=\; 0 \quad ,$$

or, using (5.1),

$$\omega\left(\frac{dB}{dt}\,v,\ v\right) + \omega([B,A]v,v) \;=\; 0 \quad ,$$

or

$$\frac{dB}{dt} \dotplus [B,A] \;=\; 0 \tag{5.6}$$

Equations (5.5) then determine the conservation laws (and observables) of the linear symplectic system (5.2).

Of course, the solution of (5.6) can be immediately written down in terms of the symplectic group ϕ_t generated by A:

$$\phi_t \;=\; e^{At} \tag{5.7}$$

$$B(t) \;=\; e^{At}B(0)e^{-At} \tag{5.8}$$

We see that the action of

$$\phi_t(\gamma)$$

on observables is essentially the <u>adjoint action</u> of the symplectic group on its Lie algebra.

We see also that if γ_1, γ_2 are conservation laws of type (5.2), determined by B_1, B_2, then we can define the Poisson bracket

$$\{\gamma_1, \gamma_2\} \quad ,$$

and it is associated with

$$[B_1, B_2] \quad .$$

Theorem 5.1. The quadratic conservation laws of a linear symplectic dynamical system form a Lie algebra (under Poisson bracket) which is isomorphic to the Lie algebra of the Lie group of symplectic automorphisms of V.

6. LINEAR UNITARY DYNAMICAL SYSTEMS

The procedure of Section 5 can, of course, be carried out for any Lie group of automorphisms of V. However, only certain types have an obvious physical motivation. I shall now describe the motivation that might lead to choosing the unitary subgroup of the symplectic group.

Keep the notation of Section 5. By considering matrices with respect to a suitable basis for V (in which ω takes its "canonical form"), the group of symplectic automorphisms of V becomes the matrix group

$$Sp(n,R)$$

($\dim V = 2n$). Suppose given a dynamical system of type (5.2). A reasonably physical condition to put on it is that:

$$\text{All solutions of (5.2) are bounded as } t \to \pm\infty \tag{6.1}$$

If one requires that this condition be satisfied, the group $t \to \phi_t$ lies in a compact subgroup of $Sp(n,R)$.

Let $U(n)$ be the subgroup of $Sp(n,R)$ consisting of the $n \times n$ complex unitary matrices regarded as $2n \times 2n$ real matrices. We now recall some facts of Lie group theory. (See Helgason [1].)

> $U(n)$ is a maximal compact subgroup of $Sp(n,R)$
>
> Any compact subgroup of $Sp(n,R)$ is conjugate--by an inner automorphism of $Sp(n,R)$ --to a subgroup of $U(n)$.

We can now translate what this means in terms of basis--free linear algebra terms.

Theorem 6.1. Suppose all solutions of the dynamical system (5.2) are bounded as $t \to \pm\infty$. Then, there is a linear map $J: V \to V$ such that:

$$AJ = JA \tag{6.2}$$

$$\omega(Jv_1, v_2) = \omega(Jv_2, v_1) \tag{6.3}$$

for $v_1, v_2 \in V$

$$J^2 = -1 \tag{6.4}$$

$$\omega(Jv,v) < 0 \tag{6.5}$$

for $v \in V - (0)$.

Remark. It is standard lore of differential geometry that this is the way to consider "complex" vector spaces from the real point of view. In other words, a real vector space V is isomorphic to a complex one, with the scalar field cut down from C to R, if and only if there is an R-linear map $J: V \to V$ such that

$$J^2 = -1 .$$

In terms of the isomorphism, J becomes $\sqrt{-1}$. Another way of putting this is to say that, given a J satisfying $J^2 = -1$, we can make V into a vector space, by defining the action of C as follows:

$$(a + bi)(v) = av + bJ(v) \tag{6.6}$$

for $a,b \in R$.

One of the prime mathematical speculative questions about quantum mechanics is: Why do the complex numbers enter, when the reals are perfectly adequate for classical mechanics? Here we see the germ of a possible "physical" or "geometric" reason for this--"boundedness" or "stability" of the dynamical system (5.2) enforces (via Cartan's maximal compact subgroup theorem) the existence of the complex structure. Of course, there are no such arguments available for infinite dimensional systems, but perhaps the fact is worth pointing out anyway.

Having restricted the dynamical system (5.2) to satisfy (6.2) - (6.5) (and these conditions are also sufficient for boundedness of all solutions as $t \to \pm\infty$), we can consider the conservation laws:

$$\gamma(v,y) = \omega(B(t)v, v) \quad , \tag{6.6}$$

with

$$\frac{dB}{dt} + [B,A] = 0 \quad . \tag{6.7}$$

Again, it is reasonable to use the special feature, the existence of J, i.e., the complex vector space structure for V, to cut down the set of B's. A natural condition to impose is that B also preserves the complex structure on V, i.e., that

$$BJ = JB \quad . \tag{6.8}$$

We shall now put conditions (6.2) - (6.5), (6.8) in a more familiar form.

Define a map

$$(v_1,v_2) \to \langle v_1,v_2\rangle \in R$$

by the formula:

$$\langle v_1, v_2 \rangle = -\omega(Jv_1, v_2) \tag{6.9}$$

for $v_1, v_2 \in V$.

<u>Exercise</u>. Show that $\langle\ ,\ \rangle$, defined by (6.9) is a positive-definite symmetric, bilinear form on V.

Now set:

$$(v_1, v_2) = \langle v_1, v_2 \rangle + i\omega(v_1, v_2) \tag{6.10}$$

$$(v_1, v_2) \to C \text{ is an R-bilinear form}$$

$$\begin{aligned} (Jv_1, v_2) &= \langle Jv_1, v_2 \rangle + i\omega(Jv_1, v_2) \\ &= -\omega(-v_1, v_2) - i\langle v_1, v_2 \rangle \\ &= -i(v_1, v_2) \end{aligned} \tag{6.11}$$

$$\begin{aligned} (v_2, v_1) &= \langle v_1, v_2 \rangle - i\omega(v_1, v_2) \\ &= \overline{(v_1, v_2)} \end{aligned} \tag{6.12}$$

Relations (6.11) - (6.12) should be formulas; they say that the inner product

$$(\ ,\ )$$

defines V as a <u>complex</u> (finite dimensional) <u>Hilbert space</u>. We can now interpret the condition that A and B must satisfy:

Exercise. Show that a linear map $A: V \to V$ is an infinitesimal symplectic automorphism, and satisfies (6.2), i.e., is C-linear if and only if

$$JA$$

is self-adjoint, i.e.,

$$(v_1, JAv_2) = (JAv_1, v_2)$$

for $v_1, v_2 \in V$.

Let us now invert this order of deduction and state directly the sort of dynamical system-state-conservation laws we will be considering from now on.

V is a complex vector space with a Hilbert space inner product

$$(v_1, v_2)$$

The states are orbits

$$t \to \phi_t(v)$$

of a one-parameter group (ϕ_t) of unitary operators on the Hilbert space V. Its infinitesimal generator

$$A: V \to V$$

is a skew self-adjoint, complex-linear map

The conservation laws are maps

$$(v,t) \to (B(t)(v),\ v) = \gamma(v,t)$$

where each $B(t)$ is a self-adjoint, C-linear map $V \to V$ such that

$$\frac{d}{dt} B + [B,A] = 0$$

The Lie algebra structure on conservation laws is given by:

$$\{B_1,B_2\} = (\text{constant})i(B_1B_2 - B_2B_1)$$

7. THE STANDARD MATHEMATICAL MODEL FOR QUANTUM MECHANICS

My approach to quantum mechanics will basically follow the "mathematical model" philosophy, despite the fact that I have stated earlier that I feel it is faulty since it is very difficult to understand the *physical* reasons for the introduction of the formalism of Hilbert spaces, self-adjoint operators, etc. However, this standard mathematical way of developing quantum mechanics gives such a useful and widely applicable picture of the physical world (or at least one side of it) that an approach that tried to do without it would be, in the present state of science, a gross violation of Occam's

Razor. I believe that quantum mechanics will eventually give way to something more plausible--at least for subnuclear phenomena--but in the meantime, we must continue to work with what we have. (For example, it would be highly desirable to add some typically "non-linear" or "global-topological" properties to quantum mechanics--or perhaps replace it altogether--in order to get a formalism which allows "elementary particles" more naturally.) Often mathematicians say--all too glibly and arrogantly--that we must do away with quantum mechanics, but the trouble is that throwing away something does not give one anything--quantum mechanics has been so uniquely successful in such a broad context of physical and chemical problems that it is easy to understand why physicists are not eager to do away with it! (Mathematicians can notice a similar phenomenon, closer to home, in the challenge by those who want to throw away non-constructive methods in the foundations of mathematics. In fact, there are broad similarities in my mind between the discussions of the foundations of quantum mechanics and the foundations of mathematics. The paradox of Schrödinger's Cat is not that different in flavor from Russel's Paradox!)

My personal opinion is that the greatest need is to push as hard as possible on the mathematics needed to make definitive mathematical deductions from the standard framework or some modification of it. I have a historical analogy in mind--the way Newtonian mechanics gave way to relativistic. Here, the two hundred years in the mathematics and physics of Newtonian mechanics--plus, of course, the discovery of a new physical phenomenon, electromagnetism--fueled the "organic" evolution of one into the other. The theoretical physicists are ever ingenious at constructing new models (without having the mathematical force or ingenuity to deduce very much from them!) and the experimentalists are giving us new experimental facts and particles. If only there were comparable programs on the mathematical side!

In order to present this standard framework for quantum mechanics, let H be a complex Hilbert space. In order to be as unambiguous as possible, I will use the mathematician's version of this. (There are some differences in this between mathematics and physics, both in notation and substance. I have tried to take into account and mediate between these differences in my own books; to do so here with any degree of substance would take up too much space.)

The best books by far for our purposes on functional analysis, and the one I will roughly follow, is Riesz-Nagy. It is far better than one published recently by two authors with the rather pretentious term "Modern" in its title! I will assume that the reader is familiar with such standard Hilbert spaces as symmetric-self-adjoint-unitary operators, Stone's theorem on one parameter unitary groups, etc. Thus, H is a complex vector space with elements denoted by $\psi_1, \psi_2, \ldots$ (possibly infinite dimensional), with a complex valued inner product

$$(\psi_1,\psi_2) \quad ,$$

satisfying (in addition to R-bilinearity),

$$(\psi_1,\psi_2) = \overline{(\psi_2,\psi_1)}$$

$$(\psi,\psi) > 0 \quad \text{if} \quad \psi \neq 0$$

$$(c\psi_1,\psi_2) = c(\psi_1,\psi_2)$$

for $c \in C$.

In addition, we suppose that the metric space defined by the norm

$$||\psi|| = (\psi,\psi)^{1/2}$$

is complete, in the functional analysis sense, i.e., "Cauchy sequences converge".

Definition. A quantum mechanical dynamical system is a pair $(H, \{\phi_t\})$ consisting of a Hilbert space H, and a one-parameter group

$$t \to \phi_t$$

of unitary automorphisms of H.

Of course, one feature introduced by the infinite dimensionality and completeness of H is that interesting classes of linear operators may not be everywhere defined.

Definition. A linear operator on H is a pair

$$(H_0, B)$$

consisting of a linear subspace $H_0 \subset H$ and a linear map

$$B: H_0 \to H_0$$

such that H_0 is dense in H. (H_0, B) is said to be symmetric if:

$$(\psi_1, B\psi_2) = (B\psi_1, \psi_2)$$

for $\psi_1, \psi_2 \in H_0$.

H_0 is said to be the domain of the operator.

Let (H_0, B) be such a symmetric operator, and let (ϕ_t) be a quantum mechanical dynamical system. They are said to be compatible if:

$$\phi_t(H_0) \subset H_0$$

for all t.

If this condition is satisfied, one can define a conservation law for the system as usual:

$$\gamma(\psi, t) = (B(t)\psi, \psi) \quad ,$$

with

$$B(t) = \phi_t B \phi_{-t}$$

Of course, one major difference between this infinite dimensional and the previously treated finite dimensional situation is that such conservation laws are not defined everywhere, but only on $H_0 \times T$. This requires us to be careful in defining an appropriate Poisson bracket structure for the observables. Here is how to do it in a situation where the domains coincide. (We will not consider more general cases here.)

<u>Definition</u>. Let B_1, B_2 be two symmetric linear operators with the same domain H_0 which are compatible with the dynamical system. Let γ_1, γ_2 be the corresponding conservation laws. Then,

$$\{\gamma_1, \gamma_2\} = \gamma_3 \quad ,$$

where γ_3 is the conservation law corresponding to the operator

$$B_3 = \frac{i}{\hbar} (B_1 B_2 - B_2 B_1)$$

($\hbar$ is, for the moment, just a real constant which is fixed throughout. Think of it as put in because it is allowed mathematically, and provides an additional flexibility to the formalism. Of course, it will be ultimately related to "Planck's Constant".)

Definition. A conservation law structure for a quantum mechanical dynamical system $(H, (\phi_t))$ is a collection of conservation laws, all with a common domain, such that the sum and Poisson bracket of two such conservation laws again lies in the collection.

We can rephrase this slightly as follows: A collection of symmetric operators on H (with common domain, H_0, which is invariant under the dynamical group) which form a Lie algebra under $i \times$ (commutator). In turn, such a collection of conservation laws determines a collection of observables (i.e., real-valued functions on states), together with a Lie algebra structure on observables inherited from the Lie algebra structure on the conservation laws.

Why should this particular choice of dynamical equations-states-conservation laws-states be so useful or natural? Here, I have two main comments in mind; one physical, the other mathematical (or, rather, Lie group-theoretical). First, mathematicians were accustomed to using symmetric operators in setting up "eigenvalue problems" for differential equations, integral equations, etc., even before the discovery of quantum mechanics. (Of course, it is a remarkable and fortuitous coincidence (?) in the history of science that this theory was essentially completed (largely by Hilbert and those he influenced or trained) just in time to be available for quantum mechanics.) In particular, eigenvalues occur in "discrete" and "continuous" families. Now, recall that Planck introduced "quantum" ideas into physics precisely in order to introduce "discreteness" in place of the "continuity" that was inevitably present in classical physics. This "eigenvalue" idea enabled the founders of quantum mechanics (who, it must be recalled, largely received their mathematical training in Germany under Hilbert's direct or indirect influence, and who were, as a group--with Heisenberg the great exception--much more familiar with the mathematics of their day than physicists are today) to allow such "discreteness" while keeping "continuity" as an underlying reality. The second Lie group theoretic feature that appeals to my mind--and of course that ties directly into the main theme of this book--is that Lie algebras of symmetric operators appear naturally as the *infinitesimal version* of *unitary representations of Lie groups*. To my mind this gives an "explanation" (which I alluded to in the introduction) why Lie group theory plays such an important role in the physical disciplines where quantum mechanics plays the basic role. Further, the "conservation laws" may be expected to play a key role in the way things *interact*. (Recall from elementary mechanics law "conservation of momentum" serves to determine the motion of two colliding particles, e.g., billiard balls.) Thus, again this is a *mathematical* clue to why certain Lie groups seem to play a role in the study of *interactions* of elementary particles. (Of course, presumably an "influence" of a Lie group in this way is compatible with *many* dynamical mechanisms for interaction, just as "symmetry" ideas can manifest themselves in diverse ways in art. It is for this reason that I have been very skeptical of elementary particles theories which expect too much from Lie group theory. I do believe, however, that <u>geometry</u> can play a key role in the creation of new physical concepts. Presumably then, the <u>combination</u> of Lie group and geometric reasoning is most appropriate for elementary particle theories--and for speculation about "fundamental physics", in general. Unfortunately, very few physicists know <u>both</u> geometry and Lie group theory well enough to use them <u>together</u> in physics.)

Having adopted such a "Hilbert space" setting for the development of quantum mechanics, one must ask: What is the <u>physical meaning</u> of the special structures we have postulated, such as the identification of "states" with elements of the Hilbert space, of observables with symmetric operators, etc.? To a large extent this is still a mystery and a source of great debate and controversy in the philosophy of science. There is something called the "Copenhagen school" of interpretation of quantum mechanics, founded by Niels Bohr, that I do not want to describe in detail. For the previous generation of physicists it was taken as the ultimate truth (or dogma?) (with the great exception of Einstein, of course), but is regarded with more skepicism today, at least by those who think about such things. (Surprisingly few physicists, by the way, actually do think about it very much. Most accept and work with it in the same "pragmatic" way that mathematicians use the Axiom of Choice. I am indebted to Abner Shimony of Boston University, one of the few physicists I have met who thinks deeply about the subject. He assures me that the Copenhagen interpretation is very much open to question and sure to be replaced by something less arbitrary some day.)

For the mathematician, there is above all the question: Why a <u>Hilbert space</u>. After all, he can understand Kronecker saying that God created integers, but why should He have so intimately concerned Himself with a Hilbert space? Now, this is <u>related</u> (but probably not necessarily causally linked) to the other controversial features of quantum mechanics--the replacement of the "certainty" of measurement in classical mechanics by a "probability" distribution. (Here I allude to another vast area--called by physicists the Theory of Measurement--that I will not go into, for lack of competence and comprehension, if nothing else.) Here, I shall only explain the mathematical mechanism, which is certainly very beautiful.

Let H be a Hilbert space. Suppose that (H_0, B) is a symmetric operator on H. Suppose further that it is actually <u>self-adjoint</u> in the technical sense this term is used in functional analysis. It is known that H is then unitary isomorphic to the Hilbert space of square integrable, measureable, complex-valued functions on a space Y, with a measure dy such that B goes over into multiplication by a measureable real valued function

$$y \to b(y) \quad .$$

Consider an element ψ of H. Normalize it so as to be of norm one, i.e.,

$$(\psi,\psi) = 1 \quad .$$

It goes over to a complex-valued function

$$y \to \psi(y)$$

on Y such that:

$$\int_Y \psi(y)\overline{\psi(y)}\, dy = 1 \quad .$$

In particular,

$$|\psi(y)|^2\, dy$$

is a <u>probability measure</u> on Y. Thus, b becomes a <u>random variable</u> with respect to this probability measure. Its "probabilit distribution" is a probability measure on the real line, defined as follows.

> The measure of an interval $[a,b] \subset R$ is the measure of $b^{-1}([a,b])$, i.e.,
>
> $$\int_{b^{-1}([a,b])} \psi(y)\overline{\psi(y)}\, dy$$

Thus, the pair (B,ψ) defines a probability measure on the real line. The average value of this measure represents the value of the observable determined by B on the state determined by ψ. This probability distribution may be defined as the "result of measurement of the observable B in the state ψ." Thus, we see that our general framework described in Section 2 for a "conservation law-observable-state" structure to dynamical systems should be extended for quantum systems as follows:

> In quantum mechanics, observables define maps
>
> (states) → (probability measures)

Remark. In case the reader wonders why the probability measure on the real line is independent of the choice of spectral resolution for B, he should notice that it can be defined "intrinsically" within the Hilbert space structure. In particular, its moments are determined by the Hilbert space structure. Its average value is:

$$\int_Y b(y)|\psi(y)|^2\, dy = (\psi, B\psi)$$

The second moment is

$$\int_Y b(y)^2 |\psi(y)|^2 \, dy = (\psi, B^2 y) \quad ,$$

etc.

It is well-known that these moments determine the probability distribution. Then, perhaps, an alternative would be to postulate a "quantum observable" as a map

$$(\text{states}) \to (\text{sequences of real numbers}) \quad ,$$

From Lie group point of view, it might be equally interesting to associate with (ψ, B) the "matrix element of the one-parameter group generated by B," namely, the following function of t

$$t \to (\psi, e^{iBt}\psi)$$

It is a positive-definite function of t. Thus, an alternative for a "quantum observable" is a map

$$(\text{states}) \to (\text{positive-definite function on the real line})$$

Finally, I want to briefly mention another mathematical feature of the Hilbert space setting for quantum mechanics. (Notice how incredibly rich in diverse mathematical ideas quantum mechanics actually is. Unfortunately, no one has ever systematically tracked down all these by-ways and put them all together!) As we have seen, the observables are symmetric operators. Now, modulo domain questions (which we, in effect, ignore), this admits two algebraic structures:

$$\{B_1,B_2\} = \frac{i}{\hbar}(B_1B_2 - B_2B_1)$$

$$[B_1,B_2]_+ = B_1B_2 + B_2B_1$$

The second--called the anti-commutative or Jordan bracket--leads to an algebraic structure called a Jordan Algebra. (In fact, P. Jordan was a physicist active in the development of quantum mechanics in the 1920's and 1930's, and he was led to their first systematic identification and study because of their obvious relevance.) They are non-associative algebras, and satisfy an identity analogous to the Jacobi identity that characterizes Lie algebras. Since Jordan's work, they have been extensively studied by algebraists, independently of physicists, much as Lie algebras were studied independently of their genesis in terms of group theory.

In summary, in this section I have briefly indicated the incredibly rich grab-bag of mathematical ideas inherent in the standard Hilbert space formulation of quantum mechanics. What has been put to the side is the question of how one chooses the dynamical system and the Lie algebra of observables in order to describe concrete and important physical systems. After all, Newtonian physics was successful not primarily because it was a beautiful mathematical structure and led to interesting new mathematics (although that played a role!) but because it unified astronomical and earth-bound mechanical phenomena. Of course, to a large extent, the derivation of the physical consequences of a mathematical model is best left to the physicists (providing he adequately understands the mathematical foundations of the model--"understanding" being taken in a rather loose sense). Our remaining job, as mathematicians, is to build a bridge from the abstract picture of the Hilbert space model for quantum mechanics to the more concrete tools, such as the Schrödinger equation. This involves, as we shall now see, being more specific about how we choose the observables and the states.

8. THE SCHRÖDINGER AND HEISENBERG EQUATIONS OF MOTION AS A CONSEQUENCE OF A CHOICE OF LIE ALGEBRA FOR THE OBSERVABLES

In previous sections I have emphasized a conception of "conservation laws-observables-states" that would apply to a diversity and generality of dynamical systems. For the remainder of this work I want to concentrate on two situations--classical particle mechanics from the symplectic manifold point of view (e.g., as in Abraham-Marsden [1]) and orthodox quantum mechanics. For these, it is convenient to rephrase and simplify the definitions of the basic concepts.

For classical mechanics, consider a manifold Q called <u>configuration space</u>. Let

$$M = T^*(Q) ,$$

the <u>cotangent bundle</u>. It is endowed with a standard symplectic structure, i.e., a closed two-differential form ω of maximal ranks. Let h be a real-valued C^∞ function on M, called the Hamiltonian. A vector field X is defined by means of h and ω, via the relation:

$$dh = \iota(X)(\omega) .$$

X generates a one-parameter group of symplectic automorphisms of M, which defines the dynamical system. The observables are then $C^\infty(M)$ (or a linear subspace), made into a Lie algebra via the Poisson bracket:

$$\{f_1,f_2\} = -X_1(f_2) ,$$

where

$$df_1 = \iota(X_1)(\omega) .$$

The states are elements of M. For $p \in M$, $f \in C^\infty(M)$, the values of the observable f on the state p (to be correlated with physical measurements in some way) is just

$$f(p) .$$

The simplest situation--and the only one we will really consider--is that where Q is R, with coordinate x. The dual coordinate to $T^*(R) \equiv M$ is then denoted by y, so that:

$$\omega = dy \wedge dx .$$

The Hamiltonian h is then a function (x,y). The dynamical system is that defined by the Hamilton equation:

$$\frac{dx}{dt} = \frac{\partial h}{\partial y}(x(t),y(t))$$

$$\frac{dy}{dx} = -\frac{\partial h}{\partial x}$$

The observables are C^∞ real-valued functions $(x,y) \to f(x,y)$. The Lie algebra structure on observables is the classical Poisson bracket:

$$\{f_1,f_2\} = \frac{\partial f_1}{\partial y}\frac{\partial f_2}{\partial x} - \frac{\partial f_2}{\partial y}\frac{\partial f_1}{\partial x}$$

Now turn to quantum mechanics. The _state space_ is the set of unit length vectors in a Hilbert space H. A _dynamical system_ is a one-parameter group

$$t \to \phi_t$$

of unitary transformations on H. An _observable_ is a symmetric operator. The Lie algebra structure on observables is

$$(B_1,B_2) \to \frac{i}{\hbar}(B_1B_2 - B_1B_2) \quad ,$$

(at least modulo the usual problems with domains).

A typical physical problem in quantum mechanics is to _construct a quantum mechanical "system"_ (i.e., the dynamics plus a suitable set of observables) which is _to describe a physical system whose classical mechanical properties are considered as_ known. (This is the problem which can loosely be described as "quantizing" a classical mechanical system. Actually, it is usually more natural mathematically to go in reverse order--from quantum mechanical properties, derive the classical. Thus, perhaps it is unfortunate that we come into physics with our intuition already shaped by "classical" ideas.)

Now, it was historically a very difficult and painful task to understand on a rational and consistent basis what the rules were relating quantum and classical properties. The landmarks of this groping period (1900-1924) was the "Bohr-Sommerfeld quantization rules". A key observation in understanding these was Ehrenfest's, that this involved "quantizing" (i.e., assigning discrete values to) what are classically called adiabatic invariants. Now, these are probably related (I am not sure exactly how) to what Poincaré and Cartan called "invariants integraux", which, in turn, are closely related to the symplectic structure of classical mechanics. The next step was due to de Broglie Heisenberg and Schrödinger--here the symplectic structure (via the relation to "Hamilton-Jacobi theory") came closer to the foreground. Finally, Dirac completely clarified the situation (or, at least, as far as it has been clarified, to this day) by proposing a rule that went something like this:

> Choose the quantum observables to be in correspondence with the classical observables in such a way that the Lie algebra structure on quantum observables

> (i.e., basically the operator commutor) corresponds to Poisson bracket of classical observables.

Now, we mathematicians would describe this in the following way:

> Choose the quantum observables so that they form a representation (by symmetric operators) of the Lie algebra (under Poisson bracket) of the classical observables.

Dirac and his successors among the physicists were always vague (purposely and wisely so, it turned out) about which Lie algebras of classical observables this was supposed to apply to. The question was never treated in any reasonable mathematical way until 1951, by L. van Hove [1], who showed that one could _not_ do this is a reasonable way by choosing as classical observables _all_ functions of (x,y), even all _polynomials_. Work since then, and my own speculations, lead me to believe that there is _no_ reasonable way to _uniquely_ "quantize" a classical system.

Of course, this negative general result did not constrain the physicists from using Dirac's idea to set up quantum systems which seemed to be an adequate version of the classical ones. What they did in practice was to choose not _all_ such classical observables, but a "physically reasonable" set, and represent _it_ by operators (with Poisson bracket--so far as it was defined within the set--going over to operator commutator).

For example, here is how to treat a system whose classical Hamiltonian has the following form:

$$h(x,y) = \frac{1}{2m} y^2 + V(x) \tag{8.1}$$

Physically, this is the total energy of a (Newtonian mechanics) particle of mass m moving in a conservative force field whose potential energy is $V(x)$. The relevant observables are

$$x, y \quad ,$$

i.e., "position" and "momentum". They satisfy the following Poisson bracket commutation relation:

$$\begin{aligned} \{y,x\} &= 1 \\ \{y,1\} &= 0 = \{x,1\} \quad . \end{aligned} \tag{8.2}$$

These say that the observables $(x,y,1)$ define a three dimensional nilpotent Lie algebra; it is, of course, now called the <u>Heisenberg Lie algebra</u>. Its Poisson bracket relations, with the Hamiltonian h take the following form:

$$\begin{aligned} \{y,h\} &= V'(x) \\ \{x,h\} &= -\frac{1}{m} y \end{aligned} \tag{8.3}$$

Let us interpret $V'(x)$ on the right hand side, not directly as a function on M but as <u>an element of the universal enveloping</u>

algebra of the Heisenberg Lie algebra (assuming, for example, that it is a polynomial in x).

Now, we can follow Dirac's prescription and represent the classical observables by quantum observables so that as far as possible Poisson brackets go into commutators. In this case, this is easy to do. Begin with the Heisenberg algebra observables (x,y,1). We must represent them by quantum observables $\underline{x},\underline{y}$, i.e., symmetric operators on a Hilbert space H, with domain H_0, such that

$$\frac{i}{\hbar}(\underline{x}\,\underline{y} - \underline{y}\,\underline{x}) = 1 \tag{8.4}$$

($\hbar$ is a constant, physically to be identified with Planck's Constant divided by 2π). Further, let us require that $\underline{x},\underline{y}$ are self-adjoint, and that $(\underline{x},\underline{y})$ act irreducibly on H. Then, there is a theorem proved by Stone and Von Neumann--which is, in a sense, the Fundamental Theorem for this method of quantization--which asserts that $(\underline{x},\underline{y},H,H_0)$ are uniquely determined up to unitary equivalence.

Having chosen one such representation of the Heisenberg algebra and observables, we must now complete the description of quantum-mechanical systems by prescribing the dynamics. This can be done by choosing a self-adjoint operator $\underline{h}$ on H whose domain contains H_0, with the dynamical unitary group defined as

$$\phi_t = \exp\left(\frac{it}{\hbar}\,\underline{h}\right) \tag{8.6}$$

We of course also require that

$$\phi_t(H_0) \subset H_0 \quad . \tag{8.7}$$

This, of course, in no way fixes $\underline{h}$. To do so, let us follow Dirac again, and ask whether $\underline{h}$ can be chosen so that the Poisson bracket relations (8.3) correspond to operator commutator relations. In other words, we require that the following conditions be satisfied:

$$\begin{aligned} \frac{i}{\hbar}(\underline{y}\,\underline{h} - \underline{h}\,\underline{y}) &= V'(\underline{x}) \\ \frac{i}{\hbar}(\underline{x}\,\underline{h} - \underline{h}\,\underline{x}) &= \frac{1}{m}\underline{y} \end{aligned} \tag{8.8}$$

It is readily verified that the following formula will do this job:

$$\underline{h} = \frac{1}{m} y^2 + V(\underline{x}) \tag{8.9}$$

Thus, assuming that everything works out analytically, i.e., $\underline{h}$ generates a one-parameter unitary group $\phi_t = \exp((i/\hbar)\underline{h}t)$, (which is not at all obvious a priori but in this case does) this procedure gives a quantum mechanical system.

As I have mentioned earlier, the Heisenberg equations of motion for the observables are then immediately obtained using $\underline{h}$:

$$\frac{d}{dt}\underline{x}_t = [\underline{h},\underline{x}_t] = \frac{1}{m}\underline{y}_t$$

$$\frac{d}{dt}\underline{y}_t = [\underline{h},\underline{y}_t] = -V'(\underline{x}_t) \quad ,$$

or

$$m \frac{d^2 \underline{x}}{dt^2} = -V'(\underline{x}_t) \tag{8.10}$$

Notice that this is of the same form as Newton's equations but is a differential equation for an operator-valued function of t, rather than--as in classical mechanics--for a real valued function of t. This gives us one general insight--which is especially relevant for quantum field theory--how the differential equation structure of quantum mechanics is related to that of classical mechanics.

So far, we have worked without any specific realization of the Hilbert space H. Let us choose H as follows:

H = square integrable complex-valued functions on the configuration space Q

(Of course, to say "square integrable" means that a measure is given for Q. In fact, Q usually comes with a Riemannian metric (in Newtonian mechanics problems this is determined by the geometry of the constraints, etc.), and the natural measure is the value associated with the Riemannian structure. In this case, with Q = R, with coordinate x, it is reasonable to choose the measure as the Lebesgue measure dx on the real line.)

Now, choose

H_0 = space of rapidly decreasing (in the Schwartz distribution sense) complex-valued functions on $R(\equiv Q)$.

$$\begin{aligned} \underline{x} &= \text{multiplication by } x \\ \underline{y} &= \frac{\hbar}{i}\frac{d}{dx} \end{aligned} \tag{8.11}$$

With this choice (which, it is readily seen, does the job-- $\underline{x}$ and $\underline{y}$ are "essentially self-adjoint", i.e., self-adjoint on their closure), we can now follow (8.9) in defining $\underline{h}$:

$$\underline{h} = -\frac{\hbar^2}{m}\frac{d^2}{dx^2} + V(x)$$

Then, the time-evolution of states

$$\psi_t = \exp \frac{i}{\hbar}\underline{h}t(\psi) \quad ,$$

is determined by the following differential equation:

$$\frac{\partial \psi_t}{\partial t} = \frac{i}{\hbar}\underline{h}(\psi_t)$$

or

$$\frac{\hbar}{i}\frac{\partial \psi}{\partial t} = -\frac{\hbar^2}{m}\frac{\partial^2 \psi}{\partial x^2} + V(x)\psi \tag{8.12}$$

(8.12) is, of course, <u>Schrödinger's equation</u>, and is the most "practical" form of quantum mechanics. The reason that quantum mechanics had the immediate success that it did was that it could be solved (in favorable cases) by the familiar separation-of-variable methods of classical mathematical physics, and the systems needed to make sense of atomic physics and chemistry (say, the hydrogen atom, harmonic oscillator) could be solved explicitly. Nowadays, an even wider variety of systems can be treated numerically using computers.

There is another way that (8.12) can be interpreted. Write down the total energy equation of Newtonian mechanics

$$E = \frac{1}{2m} y^2 + V(x) \quad .$$

Make the "operator" substitutions:

$$E \to \frac{\hbar}{i} \frac{\partial}{\partial t}$$

$$y \to \frac{\hbar}{i} \frac{\partial}{\partial x} \tag{8.13}$$

$$x \to \text{multiplication by } x \quad .$$

The Schrödinger equation (8.12) results!

Of course, since this procedure is not necessarily consistent, from a "higher" mathematical point of view it should be regarded as a speculation, to be tested for its success in making physical predictions. In other words, it is a "mathematical model" for certain types of physics. (This tentative approach also has the virtue of suggesting a bit of humility about whether the same techniques can be expected to work at the subnuclear level!)

For example, suppose the classical Hamiltonian turned out to be one of the following forms:

$$h = y^3 x + V(x) \quad ,$$

$$\sqrt{1 - xy} \quad ,$$

$$\sin\,(x\sqrt{y}) \quad .$$

What would we do to construct a quantum system? Now, the work up to now does suggest a procedure, which can be described as follows:

> If $f_1(x,y)$, $f_2(x,y)$ are two classical observables to which we have already "successfully" assigned operators $\underline{f}_1, \underline{f}_2$, assign to the product $f_1 f_2$ the "symmetrized" operator
>
> $$\frac{1}{2}\,(\underline{f}_1\underline{f}_2 + \underline{f}_2\underline{f}_1)$$

Thus, one might attempt to "quantize" an observable like

$$\sqrt{1 - xy}$$

by expanding it in a power series

$$\sqrt{1 - xy} = 1 - \frac{1}{2}\,(xy) + (\cdot)(x^2y^2) + \cdots \tag{8.14}$$

$$(xy) \rightarrow \frac{1}{2}\left(x\,\frac{\hbar}{i}\,\frac{\partial}{\partial x} + \frac{\hbar}{i}\,\frac{\partial}{\partial x}\,x\right) = \frac{\hbar}{i}\left(\psi\frac{\partial}{\partial x} + \frac{1}{2}\right) \equiv (\underline{\underline{xy}})$$

$$x^2y^2 \rightarrow ((\underline{xy}))^2$$

etc. Then, substitute these differential operators into the power series (8.14) and hope that the resulting operator can be made to make sufficient sense to deduce something useful from. (For example, the theory of pseudo differential operators is appropriate, i.e., one might try to prove that this sum is such an object.)

These "rules" for assigning operators to functions of x,y are called the Schrödinger quantization rules. In certain cases they are successful in also satisfying the Dirac conditions: Poisson brackets go into operator commutators. When this might be expected to happen is not at all understood (except for, typically, van Hove's negative result that it is not to be expected for all polynomials in x,y). When it does happen, it usually turns out to be very significant for both mathematics and physics.

For example, everything works fine for quadratic polynomials in x,y:

$$x^2 \to (\text{multiplication by } x^2)$$

$$xy \to \frac{\hbar}{i}\left(x \frac{\partial}{\partial x} + \frac{1}{2}\right)$$

$$y^2 \to -\hbar^2 \frac{\partial}{\partial x^2}$$

The Lie algebra of these observables is that of $Sp(1,R)$. The corresponding group representation is the metaplectic representation, whose study forms such a large part of the mathematical substance of Professor Wallach's book. The corresponding physical model is the <u>harmonic oscillator</u>. Similarly, for the hydrogen atom there is an "accident" of this type.

9. FURTHER REMARKS ABOUT THE PROBLEM OF QUANTIZATION OF A CLASSICAL MECHANICAL SYSTEM

I have already emphasized that there is no answer known that is satisfactory (at least from the physical point of view) to the problem of assigning a "quantum" system to a "classical" one. What seems most reasonable (and important!) as a program of research in the near future (which, at the glacial slowness of progress in mathematical physics, might be a long time!) is to break up this "big" problem into subproblems that can be treated more definitively and that have a mathematical or physical interest. To give the reader an idea what I mean, I will describe certain areas of this type which I believe are interesting.

First, let us abstract from the simple situation treated in Section 8 certain general features. We are given a symplectic manifold M. $C^\infty(M)$ has a Lie algebra structure, via Poisson bracket. Let $\underset{\sim}{G}$ be a finite dimensional Lie subalgebra of $C^\infty(M)$. Suppose our classical Hamiltonian is h, which does not belong to $\underset{\sim}{G}$. Our problem is to assign to

$$(\underset{\sim}{G}, h, M)$$

a quantum mechanical system.

In practice, one often finds that h is built up from polynomials (in the function sense) from elements of $\underset{\sim}{G}$. Let us suppose this is so. One might then attempt to construct the quantum mechanical system in the following way.

Let G be the simply connected Lie group whose Lie algebra is $\underset{\sim}{G}$. Let

$$\rho: G \to U(H)$$

be a linear representation of G be unitary operator on a Hilbert space H. One can then pick a domain $H_0 \subset H$, and an infinitesimal representation $\rho(\underset{\sim}{G})$ of $\underset{\sim}{G}$ by skew-adjoint operators on the domain H_0, such that each $\rho(X)$, for $X \in \underset{\sim}{G}$, is <u>essentially self adjoint</u>. (For example, the "C^∞ vectors" or the "analytic vectors" will do.) Assign to the classical observables in $\underset{\sim}{G}$ essentially self-adjoint operators (i.e., quantum observables) on H_0 as follows:

$$X \to \frac{\hbar}{i} \rho(X) \tag{9.1}$$

To construct the dynamics, <u>try</u> to assign an operator $\underline{h}$ to h, as an appropriate "polynomial" in the elements of $\rho(\underset{\sim}{G})$. I will now attempt to describe this in general terms.

Let $T(\underset{\sim}{G})$ denote the (associative) tensor algebra generated <u>by the vector space</u> $\underset{\sim}{G}$. The adjoint representation Ad $\underset{\sim}{G}$ extends to a representation in $T(\underset{\sim}{G})$:

$$\text{Ad } X(X_1 \otimes \cdots \otimes X_n) = [X,X_1] \otimes \cdots \otimes X_n + \cdots + X_1 \otimes \cdots \otimes [X,X_n] \quad (9.2)$$

for $X, X_1, \ldots, X_n \in \underset{\sim}{G}$.

We can now define an (associative) algebra homomorphism

$$\alpha \colon T(\underset{\sim}{G}) \to C^\infty(M)$$

as follows.

$$\alpha(X_1 \otimes \cdots \otimes X_n) = f_1, \ldots, f_n \quad ,$$

where $f_1, \ldots, f_n$ are the functions on M corresponding to $X_1, \ldots, X_n$. (Recall that $\underset{\sim}{G}$ is <u>given</u> as a Lie subalgebra of $C^\infty(M)$.)

Our assumption that "h is a polynomial in the elements of $\underset{\sim}{G}$" means that

$$h \in \alpha(T(\underset{\sim}{G})) \quad .$$

We can also extend ρ to define an associative algebra homomorphism

$$\rho \colon T(\underset{\sim}{G}) \to (\text{operators on } H)$$

as usual.

$$\rho(X_1 \otimes \cdots \otimes X_n) = \rho(X_1) \cdots \rho(X_n) \left(\frac{\hbar}{i}\right)^n$$

We can then propose to define the "quantum mechanics" by an element $h' \in \alpha^{-1}(h)$, such that:

$$\underline{h} \equiv \rho(h')$$

is symmetric. If we are lucky, there is only one such element! One could conceive of other rules for determining the quantum Hamiltonian--this one seems to work in important cases.

For example, consider the hydrogen atom. Follow the treatment in my "Lie Groups for Physicists", pp. 144-146. Here,

$$Q = R^3 , \qquad M \equiv T^*(Q) = R^6 .$$

Choose indices as:

$$1 \leq i,j,k \leq 3 ,$$

and the summation convention. Let

$$(x_i)$$

be coordinates for Q, (x_i, y_i) be canonical coordinates for M, i.e., the symplectic form is

$$\omega = dy_i \wedge dx_i$$

The Hamiltonian for the classical hydrogen atom is (in appropriately normalized units):

$$h = \frac{1}{2} p_i p_i - \frac{1}{r}$$

with $r = x_i x_i$. Set:

$$f_{ij} = x_i y_j - x_j y_i ,$$

$$f_i = \frac{x_i}{r} - x_i(y_j y_j) + (x_j y_j) y_i$$

The following Poisson bracket relations are satisfied:

$$0 = \{f_i, h\}$$

$$\{f_i, f_j\} = 2h f_{ij}$$

$$\{f_{ki}, f_j\} = (\delta_{ji}\delta_{k\ell} - \delta_{jk}\delta_{ij}) f_\ell$$

Restrict attention to the region of M where

$$h < 0$$

(Physically, this corresponds to the "bound state region".) Then,

$$\left\{\frac{f_i}{\sqrt{-h}}, \frac{f_j}{\sqrt{-h}}\right\} = -2f_{ij}$$

Also,

$$\frac{1}{h} = \frac{f_i}{\sqrt{-h}} \frac{f_i}{\sqrt{-h}} + f_{ij} f_{ij} \tag{9.1}$$

We can now let $\underset{\sim}{G}$ be the Lie algebra generated by the $f_i/\sqrt{-h}$ and f_{ij}. (It is isomorphic to the Lie algebra of $SO(4,R)$.) It turns out (via work of Fock and Bargmann) that the appropriate representation is that where:

H = square integrable functions on the three-sphere

$\rho(G)$ = natural "geometric" action of SO(4,R) on the functions on the three-sphere

h is defined as

$$\left(-\frac{1}{\hbar^2}(\rho(X_i)\rho(X_i)+\rho(X_{ij})\rho(X_{ij})^{-1}\right)$$

The beauty of this formula is that it can be used very easily to calculate the eigenvectors and eigenvalues of h. In fact, they are

$$-\frac{\hbar^2}{j(j+1)}, \qquad j = 1,2,\ldots,$$

where the j-th eigenspace carries (irreducibly) the (spin j) ⊗ (spi representation of SO(4,R). (Recall that SO(4,R) is locally isomorphic to SO(3,R) × SO(3,R).) This is the famous Balmer formula for the hydrogen atom; it was one of the great accomplishments of quantum mechanics to derive it on a mathematical basis. (The first such derivation, by Schrödinger, solved his equation in a brute-force way.)

This example is so beautiful that it has been a constant inspiration for physicists who hope to apply Lie group theory to physics. (Although most of this work has been in the last ten years, after the success of the Eightfold Way (i.e., SU(3)) in elementary particle physics. When I was writing "Lie Groups for

Physicists" in 1964-66, I could find no physicists who knew anything about it, and had to figure it out myself. (One of the tragedies of mathematical physics is that the work of the pioneers in quantum mechanics was not carried on by the next generation, and many of the ideas were lost.)

Obviously, there are many features to this example which one would like to understand better and more systematically. For example:

a) What are the classical mechanics system for which a method like this works?

b) Given a Lie algebra $\underset{\sim}{G}$ that can be used in this way, how are the physically relevant representations $\rho(\underset{\sim}{G})$ obtained?

c) Are there any interesting physically interesting systems with an infinite number of degrees of freedom (e.g., a "field theory") that can be obtained in this way? (This might require using representation of infinite dimensional Lie algebras, about which very little is known.) In particular, the observed physical properties of elementary particles are suggested for such an approach!

There have been many relevant fragmentary results in the physics literature in the last ten years, but not too much systematic work is available.

Bibliography

1. R. Abraham and J. Marsden, Foundations of Mechanics, W.A. Benjamin, N.Y., 1967.

1. H. Corben and P. Stehle, Classical Mechanics, Wiley, N.Y., 1960.

1. R. Feynman, Lectures on Physics, Addison-Wesley, Reading, Mass., 1963.

1. H. Goldstein, Classical Mechanics, Addison-Wesley, Reading, Mass., 1951.

1. K. Gottfried, Quantum Mechanics, W.A. Benjamin, N.Y., 1966

1. S. Helgason, Differential Geometry and Symmetric Spaces, Academic Press, N.Y., 1962.

1. R. Hermann, Remarks on the Geometric Nature of Quantum Phase Space, J. Math. Phys. 6 (1965), 1768-1771.

2. R. Hermann, Lie Groups for Physicists, W.A. Benjamin, N.Y., 1966.

3. R. Hermann, Fourier Analysis on Groups and Partial Wave Analysis, W.A. Benjamin, N.Y., 1969.

4. R. Hermann, Lie Algebras and Quantum Mechanics, W.A. Benjamin, N.Y., 1970.

5. R. Hermann, Vector Bundles in Mathematical Physics, Parts I&II, W.A. Benjamin, N.Y., 1970.

6. R. Hermann, Lectures on Mathematical Physics, Parts I&II, W.A. Benjamin, 1970 and 1972.

7. R. Hermann, Spectrum Generating Algebras in Classical Mechanics, I&II, J. Math. Phys. 13 (1972), 833, 878.

8. R. Hermann, Geometry, Physics, and Systems, M. Dekker, N.Y., 1973.

9. R. Hermann, Physical Aspects of Lie Group Theory, Univ. of Montreal Press, 1974.

10. R. Hermann, Topics in the Mathematics of Quantum Mechanics, (Interdisciplinary Mathematics, Vol. 6), Math Sci Press, Brookline, 1973.

11. R. Hermann, Gauge Fields and Cartan-Ehresmann Connections, Part A. (Interdisciplinary Mathematics, Vol. 10), Math Sci Press, Brookline, 1975.

12. R. Hermann, Higher Derivative Conservation Laws and Symplectic Manifolds, to appear, J. Math. Phys.

1. G. Mackey, The Mathematical Foundations of Quantum Mechanics, W.A. Benjamin, N.Y., 1963.

1. A. Messiah, Quantum Mechanics, North-Holland, Amsterdam, 1961.

1. J. Moyal, Quantum Mechanics as a Statistical Theory, Proc. Cambridge Philos. Soc. 45 (1949), 99-124.

1. L. O'Raifeartaigh, Unitary Representations of Lie Groups in Quantum Mechanics, in "Lecture Notes in Physics" Vol. 6, Springer-Verlag, Berlin, 1970.

1. F. Prugovecki, Quantum Mechanics in Hilbert Space, Academic Press, 1971.

1. M. Reed and B. Simon, Modern Methods of Mathematical Physics, Vols. I&II, Academic Press, N.Y.

1. F. Riesz and B. Nagy, Functional Analysis, Ungar, N.Y., 1955.

1. R. Thom Structural Stability and Morphogenesis, W.A. Benjamin, Reading, Mass., 1975.

1. L. Van Hove, Sur certaines représentations unitaries d'un groupe infini de transformations, Mem. Acad. Roy. Belg. 26, 1951.